Modeling, Simulation, and Optimization of Supercritical and Subcritical Fluid Extraction Processes

Modeling, Simulation, and Optimization of Supercritical and Subcritical Fluid Extraction Processes

Zainuddin A. Manan
Universiti Teknologi Malaysia (UTM)
Johor Bahru, Malaysia

Gholamreza Zahedi
Universiti Teknologi Malaysia (UTM)
Johor Bahru, Malaysia

and

Ana Najwa Mustapa
College of Engineering
Universiti Teknologi MARA (UiTM)
Shah Alam, Malaysia

Copyright © 2022 by the American Institute of Chemical Engineers, Inc. All rights reserved.
A Joint Publication of the American Institute of Chemical Engineers and John Wiley & Sons, Inc.

Published by John Wiley & Sons, Inc., Hoboken, New Jersey.
Published simultaneously in Canada.

All rights reserved. No part of this publication may be reproduced, stored in a retrieval system, or transmitted, in any form or by any means, electronic, mechanical, photocopying, recording or otherwise, except as permitted by law. Advice on how to obtain permission to reuse material from this title is available at http://www.wiley.com/go/permissions.

The right of Zainuddin A. Manan, Gholamreza Zahedi, and Ana Najwa Mustapa to be identified as the authors of this work has been asserted in accordance with law.

Registered Office
John Wiley & Sons, Inc., 111 River Street, Hoboken, NJ 07030, USA

Editorial Office
John Wiley & Sons, Inc., 111 River Street, Hoboken, NJ 07030, USA

For details of our global editorial offices, customer services, and more information about Wiley products visit us at www.wiley.com.

Wiley also publishes its books in a variety of electronic formats and by print-on-demand. Some content that appears in standard print versions of this book may not be available in other formats.

Limit of Liability/Disclaimer of Warranty
In view of ongoing research, equipment modifications, changes in governmental regulations, and the constant flow of information relating to the use of experimental reagents, equipment, and devices, the reader is urged to review and evaluate the information provided in the package insert or instructions for each chemical, piece of equipment, reagent, or device for, among other things, any changes in the instructions or indication of usage and for added warnings and precautions. While the publisher and authors have used their best efforts in preparing this work, they make no representations or warranties with respect to the accuracy or completeness of the contents of this work and specifically disclaim all warranties, including without limitation any implied warranties of merchantability or fitness for a particular purpose. No warranty may be created or extended by sales representatives, written sales materials, or promotional statements for this work. The fact that an organization, website, or product is referred to in this work as a citation and/or potential source of further information does not mean that the publisher and authors endorse the information or services the organization, website, or product may provide or recommendations it may make. This work is sold with the understanding that the publisher is not engaged in rendering professional services. The advice and strategies contained herein may not be suitable for your situation. You should consult with a specialist where appropriate. Further, readers should be aware that websites listed in this work may have changed or disappeared between when this work was written and when it is read. Neither the publisher nor authors shall be liable for any loss of profit or any other commercial damages, including but not limited to special, incidental, consequential, or other damages.

Library of Congress Cataloging-in-Publication Data Applied for:

ISBN: 9781118460177

Cover Design: Wiley
Cover Image: Courtesy of Zainuddin A Manan

Set in 10/13pt STIXTwoText by Straive, Pondicherry, India

10 9 8 7 6 5 4 3 2 1

Contents

Preface *xiii*
Nomenclature *xvii*

1   **Fundamentals of Supercritical and Subcritical Fluid Extraction**   *1*
1.1   Introduction   *1*
1.2   Supercritical Fluid Properties   *2*
1.3   Subcritical Condition   *3*
1.4   Physical Properties of Subcritical Fluid   *5*
1.5   Principles of Sub- and Supercritical Extraction Process   *7*
1.5.1   Solid Sample Extraction   *8*
1.5.2   Liquid Sample Extraction   *9*
1.6   Applications of SCF Extraction   *11*
1.6.1   Decaffeination of Coffee and Tea   *11*
1.6.2   Removal of FFA in Fats and Oils   *15*
1.6.3   Enrichment of Tocopherols   *17*
1.6.4   Carotenes from Crude Palm Oil and from Palm Fatty Acid Esters   *18*
1.7   Solubility of Solutes in SCFs   *18*
1.8   Solute–Solvent Compatibility   *20*
1.9   Solubility and Selectivity of Low-Volatility Organic Compounds in SCFs   *21*
1.10   Method of Solubility Measurement   *24*
1.10.1   Static Method   *24*
1.10.2   Dynamic Method   *25*
1.11   Determination of Solvent   *27*
1.11.1   Carbon Dioxide (CO_2)   *30*
1.11.2   1,1,1,2-Tetrafluoroethane (R134a) as a Solvent   *31*
1.12   Important Parameters Affecting Supercritical Extraction Process   *36*

1.12.1	Pressure and Temperature	*36*
1.12.2	Solvent Flowrate	*38*
1.12.3	Cosolvent	*39*
1.12.4	Moisture Content	*40*
1.12.5	Raw Material	*42*
1.13	Profile of Extraction Curves	*43*
1.14	Design and Scale Up	*45*
2	**Modeling and Optimization Concept**	*47*
2.1	SFE Modeling	*47*
2.1.1	Importance of Knowing the Solid Matrix and Selecting a Suitable Model	*48*
2.1.2	Different Modeling Approaches in SFE	*48*
2.1.2.1	Experimental Models	*49*
2.1.2.2	Models Which Are Based on Similarity between Heat and Mass Transfer	*49*
2.1.2.3	Models Based on Conservation Balance Equations	*49*
2.2	First Principle Modeling	*49*
2.2.1	The Equation of Continuity	*50*
2.2.2	The Equation of Motion in Terms of τ	*50*
2.2.3	The Equation of Energy in Terms of q	*52*
2.3	Hybrid Modeling or Gray Box	*53*
2.4	ANN	*55*
2.4.1	Simple Neural Network Structure	*55*
2.4.1.1	Transfer Function	*57*
2.4.1.2	Activation Functions	*57*
2.4.1.3	Learning Rules	*57*
2.4.2	Network Architecture	*58*
2.5	Fuzzy Logic	*61*
2.5.1	Boolean Logic and Fuzzy Logic	*61*
2.5.2	Fuzzy Sets	*62*
2.5.3	Membership Function	*63*
2.5.3.1	Membership Function Types	*63*
2.5.4	Fuzzy Rules	*64*
2.5.4.1	Classical Rules and Fuzzy Rules	*65*
2.5.5	Fuzzy Expert System and Fuzzy Inference	*66*
2.5.5.1	Mamdani FIS	*66*
2.5.5.1.1	Fuzzification	*66*

2.5.5.1.2	Fuzzy Logical Operation and Rule Evaluation	66
2.5.5.1.3	Implication Method	67
2.5.5.1.4	Aggregation of the Rule Outputs	67
2.5.5.1.5	Defuzzification	67
2.5.5.2	Sugeno Fuzzy Inference	67
2.6	Neuro Fuzzy	68
2.6.1	Structure of a Neuro Fuzzy System	69
2.6.2	Adaptive Neuro Fuzzy Inference System (ANFIS)	69
2.6.2.1	Learning in the ANFIS Model	71
2.7	Optimization	72
2.7.1	Traditional Optimization Methods	73
2.7.2	Evolutionary Algorithm	74
2.7.3	Simulated Annealing Algorithm	74
2.7.4	Genetic Algorithm	75
2.7.4.1	Genetic Algorithm Definitions	75
2.7.4.2	Genetic Algorithms Overview	76
2.7.4.3	Preliminary Considerations	77
2.7.4.4	Overview of Genetic Programming	78
2.7.4.5	Implementation Details	79
2.7.4.5.1	Selection Operator	79
2.7.4.5.2	Crossover Operator	79
2.7.4.5.3	Mutation Operator	79
2.7.4.6	Effects of Genetic Operators	80
2.7.4.7	The Algorithms	80
3	**Physical Properties of Palm Oil as Solute**	**83**
3.1	Introduction	83
3.2	Palm Oil Fruit	83
3.3	Palm Oil Physical and Chemical Properties	84
3.3.1	Palm Oil Triglycerides	85
3.3.2	Minor Components in Palm Oil	89
3.4	Vegetable Oil Refining	91
3.5	Conventional Palm Oil Refining Process	91
3.5.1	Chemical Refining	93
3.5.2	Physical Refining	97
3.5.3	Effect of Palm Oil Refining	98
3.6	Conclusions	100

4	**First Principle Supercritical and Subcritical Fluid Extraction Modeling** *101*
	Part I: Modeling Methodology *101*
4.1	Introduction *101*
4.2	Phase Equilibrium Modeling *101*
4.3	The Redlich–Kwong–Aspen Equation of State *102*
4.3.1	Calculations of Pure Component Parameters for the RKA-EOS *102*
4.3.2	Binary Mixture Calculations *103*
4.4	Palm Oil System Characterization *103*
4.4.1	Palm Oil Triglycerides *104*
4.4.2	Free Fatty Acids *106*
4.4.3	Palm Oil Minor Components *106*
4.5	Development of Aspen Plus® Physical Property Database for Palm Oil Components *107*
4.5.1	Vapor Pressure Estimation *107*
4.5.2	Estimation of Pure Component Critical Properties *108*
4.5.2.1	Critical Properties Estimation Using Normal Boiling Point *108*
4.5.2.2	Critical Properties Estimation Using One Vapor Pressure Point *110*
4.6	Binary Interaction Parameters Calculations *110*
4.7	Supercritical Fluid Extraction Process Development *113*
4.7.1	Hydrodynamics of Countercurrent SFE Process *113*
4.7.2	Solubility of Palm Oil in Supercritical CO_2 *115*
4.7.3	Process Modeling and Simulation *116*
4.7.3.1	Simple Countercurrent Extraction *118*
4.7.3.2	Countercurrent Extraction with External Reflux *118*
4.7.4	Process Analysis and Optimization *119*
	Part II: Results and Discussion *120*
4.8	Palm Oil Component Physical Properties *120*
4.8.1	Vapor Pressure of Palm Oil Components *120*
4.8.2	Pure Component Critical Properties *122*
4.9	Regression of Interaction Parameters for the Palm Oil Components-Supercritical CO_2 Binary System *122*
4.9.1	Binary System: Triglyceride – Supercritical CO_2 *123*
4.9.2	Binary System: Oleic Acid – Supercritical CO_2 *126*
4.9.3	Binary System: α-Tocopherol – Supercritical CO_2 *128*
4.9.4	Binary System: β-Carotene – Supercritical CO_2 *130*
4.9.5	Temperature-Dependent Interaction Parameters *131*
4.10	Phase Equilibrium Calculation for the Palm Oil–Supercritical CO_2 System *132*

4.11	Ternary System: CO_2 – Triglycerides – Free Fatty Acids	*133*
4.12	Distribution Coefficients of Palm Oil Components	*134*
4.13	Separation Factor Between Palm Oil Components	*138*
4.13.1	Separation Factor Between Fatty Acids and Triglycerides	*139*
4.13.2	Separation Factor Between Fatty Acids and α-Tocopherols	*140*
4.14	Base Case Process Simulation	*141*
4.14.1	Palm Oil Deacidification Process	*141*
4.14.1.1	Solubility of Palm Oil in Supercritical CO_2	*141*
4.14.1.2	Palm Oil Deacidification Process: Comparison to Pilot Plant Results	*142*
4.15	Conclusion	*145*
5	**Application of Other Supercritical and Subcritical Modeling Techniques**	***147***
5.1	Mass Transfer, Correlation, ANN, and Neuro Fuzzy Modeling of Sub-and Supercritical Fluid Extraction Processes	*147*
5.2	Mass Transfer Model	*148*
5.3	ANN Modeling	*153*
5.4	Neuro Fuzzy Modeling	*153*
5.5	ANFIS and Gray-box Modeling of Anise Seeds	*154*
5.6	White Box SFE Modeling of Anise	*155*
5.6.1	Gray Box Parameters	*156*
5.6.2	ANFIS	*156*
5.6.2.1	Preprocessing	*157*
5.6.3	Gray Box	*158*
5.7	Results and Discussion	*159*
5.7.1	ANFIS	*159*
5.7.2	Gray Box Modeling Results	*159*
5.7.2.1	Black Box	*159*
5.7.3	Comparison of ANFIS and Gray Box Models with ANN and White Box Models	*161*
5.8	Introduction – Statistical versus ANN Modeling	*162*
5.9	Supercritical Carbon Dioxide Extraction of *Q. infectoria* Oil	*164*
5.9.1	Materials and Methods	*165*
5.9.2	Experimental Design	*165*
5.9.3	Artificial Neural Network Modeling	*168*
5.10	Subcritical Ethanol Extraction of Java Tea Oil	*168*
5.10.1	Artificial Neural Network Modeling	*172*
5.11	SFE of Oil from Passion Fruit Seed	*173*
5.11.1	Experimental Procedures	*173*

5.11.2	RSM Statistical Modeling	*174*
5.11.3	ANN Modeling of Passion Fruit Seed Oil Extraction with Supercritical Carbon Dioxide	*176*

6 Experimental Design Concept and Notes on Sample Preparation and SFE Experiments *179*

6.1	Introduction	*179*
6.2	Experimental Design	*179*
6.3	Statistical Optimization	*180*
6.4	Optimization of Palm Oil Subcritical R134a Extraction	*182*
6.4.1	Effect of Temperature and Pressure	*184*
6.4.2	Model Fitting	*187*
6.4.3	Process Optimization	*189*
6.5	Comparison of Subcritical R134a and Supercritical CO_2 Extraction of Palm Oil	*190*
6.5.1	Extraction Performance	*191*
6.5.2	Economic Factor	*196*
6.6	Sample Pretreatment	*197*
6.6.1	Moisture Content Reduction	*198*
6.6.2	Sample Size Reduction	*199*
6.7	New Trends in Pretreatment	*200*
6.8	Optimal Pretreatment	*203*

7 Supercritical and Subcritical Optimization *205*

Part I: First Principle Optimization *205*

7.1	Introduction	*205*
7.2	Evaluation of Separation Performance	*205*
7.2.1	Effects of Temperature and Pressure	*206*
7.2.2	Effect of the Number of Stages	*207*
7.2.3	Effect of Solvent-to-Feed Ratio	*208*
7.2.4	Effect of Reflux Ratio	*209*
7.3	Parameter Optimization of Palm Oil Deacidification Process	*210*
7.3.1	Simple Countercurrent Extraction (Without Reflux)	*212*
7.3.2	Countercurrent Extraction with Reflux	*213*
7.4	Proposed Flowsheet for Palm Oil Refining Process	*215*
7.5	Conclusions	*216*

Part II: ANN, GA Statistical Optimization *217*

7.6 Introduction *217*
7.7 Traditional Optimization *217*
7.8 Nimbin Extraction Process Optimization *220*
7.9 Genetic Algorithm for Mass Transfer Correlation Development *223*
7.10 Optimizing Chamomile Extraction *225*
7.11 Statistical and ANN Optimization *227*
7.12 Conclusion *232*

Appendix A **Calculation of the Composition for Palm Oil TG (Lim et al. 2003)** *233*
Appendix B **Calculation of Distribution Coefficient and Separation Factor (Lim et al. 2003)** *235*
Appendix C **Calculation of Palm Oil Solubility in Supercritical CO_2 (Lim et al. 2003)** *237*
References *239*
Index *265*

Preface

Supercritical and subcritical fluid extraction (SFE/SCFE) technologies have become increasingly popular methods for extraction and purification of food ingredients, cosmetics, and pharmaceuticals over the last 30 years due to their unique advantages over conventional processing methods. These include low-temperature operation, inert solvent, selective separation, and the extraction of high-value product or new product with improved functional or nutritional characteristics. SFE/SCFE are also environmentally benign technologies since the processes typically generate no waste.

Supercritical fluid exhibits high-density like liquids, which contributes to greater potential for solubilization of materials, and low viscosity similar to gases, which enables its penetration into the solid. Subcritical fluid, which is also known as a high-pressure liquid, exhibits similar behavior to and can be exploited in the same manner as, supercritical fluids albeit at much lower pressure and temperature. SCFE is therefore typically classified under SFE technology. Nowadays, SFE technology is used to process hundreds of millions of pounds of coffee, tea, and hops annually, and is increasingly becoming of common use in the pharmaceuticals industry for purification and nanoparticle formation. Supercritical fluid processing is also gaining in the botanicals, vitamins, and supplements industries, where they are becoming synonymous with the highest purity and quality.

Commercial application of SFE technology has been relegated to only special applications due to high-equipment capital cost required. Besides that, conceptual development of extraction processes for natural oils, fats, cosmetic, and pharmaceuticals based on supercritical fluid technology is hindered by the lack of suitable design tools and reliable thermodynamic data and models at high pressure. Optimization of SFE processes involves a search for the optimum conditions above the critical point that is in a narrow limit and needs special care. Process simulation is used for the design, development, analysis, and optimization of technical processes and is mainly applied to chemical plants and chemical processes. Integration of optimization techniques into simulation practice, specifically into commercial software, has become nearly ubiquitous, as most discrete-event simulation packages now include some form of optimization routines. Even though modeling, simulation, and optimization tools have been widely used for design of chemical processes, their application has been very limited in vegetable oil processing particularly at supercritical conditions. In line with this limitation, there is a dearth need for literature on this topic. So far there has been no book written on modeling, simulation optimization of SFE processes in the market.

This book provides a complete guideline on tools and techniques for modeling of SFE as well as SCFE processes and phenomena and provides details for both SFE and SCFE from managing the experiments to modeling and simulation optimization. The book also includes the fundamentals of SFE as well as the necessary experimental techniques to validate the models.

The simulation optimization section includes the use of process simulators, conventional optimization techniques, and state-of-the-art genetic algorithm methods. Numerous practical examples and case studies on the application of the modeling and optimization techniques on the SFE processes are also provided. Detailed thermodynamic modeling with and without cosolvent and nonequilibrium system modeling is another feature of the book.

The book consists of seven chapters. Chapter 1 presents an overview of the field of supercritical and subcritical fluid extraction (SSFE) and their importance to food, cosmetic and pharmaceutical industries. Chapter 2 describes the concepts and methodologies for modeling, simulation, and optimization. It presents conservation laws related to SFE traditional first principle modeling and optimization techniques, as well as advanced artificial intelligence (AI) techniques such as genetic algorithm, fuzzy logic, and artificial neural network. The characteristics and physical properties of palm oil as the most referred solute in the book, and descriptions of some existing palm oil industrial processes are presented in Chapter 3. In Chapter 4, the first principle methodology is applied for modeling of properties of palm oil components, mixtures, and for the SFE of palm oil components. Modeling applications involving advanced techniques such as AI, ANN, and fuzzy logics and ANFIS are discussed in Chapter 5. Next, Chapter 6 describes experimental design concepts and procedures as well as statistical optimization techniques involving SSFE processes. Finally, optimization of SSFE using first principle modeling and other advanced techniques are presented in Chapter 7. The colored version of few figures from this chapter can be viewed on the product's page of the following website, https://www.wiley.com

Zainuddin Abdul Manan, Gholamreza Zahedi,
School of Chemical and Energy Engineering,
Faculty of Engineering, Universiti Teknologi Malaysia,
UTM Johor Bahru 81310
Johor Malaysia
Ana Najwa Mustapa
Universiti Teknologi MARA

Nomenclature

Symbol	Definition
AAD:	Average Absolute Deviation
ANFIS:	Adaptive Neuro Fuzzy Inference System
ANN:	Artificial Neural Network
Bi:	Biot number, $(k_f R_p)/D_s$
C (kg/m³):	Oil concentration in the supercritical fluid phase
$C\left(\dfrac{\text{kmol}}{\text{m}^3}\right)$:	Oil concentration in the supercritical phase
COG:	Centre of Gravity
$c_s\left(\dfrac{\text{kmol}}{\text{m}^3}\right)$:	Oil concentration at the surface of the vetiver particle
D_{ext} (m):	Diameter of extraction column
$D_l\left(\dfrac{\text{m}^2}{\text{s}}\right), D_m\left(\dfrac{\text{m}^2}{\text{s}}\right), D_s\left(\dfrac{\text{m}^2}{\text{s}}\right)$:	Axial dispersion coefficient, Molecular diffusion coefficient, Diffusivity of oil in the vetiver particle
F:	Percent of extract
FIS:	Fuzzy Inference System
FPM:	First Principle Model
GA:	Genetic Algorithm
GB:	Gray Box
K:	Extract equilibrium constant between solid and fluid phase (-)
k:	Equilibrium constant
$k_f\left(\dfrac{\text{m}}{\text{s}}\right)$:	Mass transfer coefficient

L (m):	Length of extractor
MLP:	Multilayer Perceptron
n_0 (kmol):	Initial mole of solute in the bed
NF:	Neuro-Fuzzy
OECs:	Overall Extraction Curves
P (MPa):	Pressure
PDE:	Partial differential equation
Pe_b, Pe_p:	Peclet number for the bed, $(L\,\nu)/D_l$, Peclet number for the vetiver particle, $(R_p\nu)/D_s$
Q:	Degree of Membership
$Q\left(\dfrac{\text{kg}}{\text{s}}\right)$:	Flow rate of supercritical fluid, $(\nu A\varepsilon\rho_f)$
q (kg/m³):	Oil concentration in the solid phase
Re:	Reynolds number
RMSE:	Root Mean Square Error
R_p (m):	Vetiver particle radius
r (m):	Axial coordinate in the vetiver particle
SFE:	Supercritical Fluid Extraction
Sc:	Schmidt number, $\mu_f/(\rho_f D_m)$
Sh:	Sherwood number, $(2R_p k_f)/D_m$
T:	Temperature (K)
t:	Time (s)
V (m/s):	Velocity of the fluid
WB:	White Box
x (m):	Distance of a point in bed from place of input fluid
x_0:	Initial mass fraction of extractable oil in solid phase
z:	Dimensionless axial coordinate along the bed, x/L

Subscript

a:	Apparent
b:	Bed
c:	Critical
ext:	Fluid
i:	Inter phase
p:	Particle
s:	Surface of particle
0:	At time zero

Greek Letter

$\mu \left(\dfrac{\text{kg}}{\text{m.s}}\right)$:   Supercritical fluid viscosity

μ: Membership function

ρ: Dimensionless radial coordinate in the vetiver particle, r/Rp

ρ (kg/m^3):   Density

$\rho_f \left(\dfrac{\text{kg}}{\text{m}^3}\right)$:   Supercritical fluid density

ε: Void fraction of packed bed

$\nu \left(\dfrac{\text{m}}{\text{s}}\right)$:   Interstitial fluid velocity

τ: Dimensionless time, $(t\nu)/L$

$\dfrac{\partial \rho}{\partial t}$:   Rate of increase of mass per unit volume

$-(\nabla \cdot \rho v)$:   Net rate of mass addition per unit volume by convection

ρv: Mass flux

$(\nabla \cdot \rho v)$:   Net rate of mass efflux per unit volume

$\dfrac{\rho Dv}{Dt}$:   Mass per unit volume times acceleration

$-\nabla \rho$:   Pressure force on element per unit volume

$-[\nabla \cdot \tau]$:   Viscous force on element per unit volume

$+\rho g$:   Gravitational force on element per unit volume

$-(\nabla \cdot q)$:   Rate of internal energy addition by heat conduction per unit volume

$-(\tau : \nabla v)$:   Irreversible rate of internal energy increase per unit volume by viscous dissipation

1

Fundamentals of Supercritical and Subcritical Fluid Extraction

1.1 Introduction

The behavior of supercritical fluids (SCF) was first observed and reported by Charles Cagniard de la Tour in 1822. From his early experiments, the critical point of a fluid was first discovered, and the unique properties of the SCF phase were observed. Many decades later, the power of SCFs to act as solvents for substances in solid matrices was demonstrated by Hannay and Hogarth in 1879. The group of scientists found that increasing pressure tended to dissolve solutes, while decreasing pressure caused the dissolved materials to precipitate like snow. These observations were fundamental to the understanding of the SCF extraction (SFE) technology. Zosel (1974) first demonstrated the decaffeination of coffee using SC fluids. Since then, numerous scientific works have emerged from a wide range of applications including food industry, polymers, and pharmaceuticals. The use of a SCF as a solvent to selectively extract a substance is known as SFE. The extracted material is typically recovered simply by reducing the temperature or pressure of the SCF, thereby allowing the fluid to evaporate.

SFE technology has received much attention over the last two decades. Factors like the high cost of organic solvent, increasingly stringent environmental regulations along with new requirements from the medical and food industries for ultrapure and high-quality products have driven the development of new and clean technologies for the recovery of substances. These special needs have resulted in SFE technology applications in the food, aroma, and waste treatment industries during the 1980s. In the food and pharmaceutical industries, for example, SFE has been applied for the extraction of flavors from hops, cholesterol, and fat from eggs, nicotine from tobacco, acetone from antibiotics, and for coffee and tea decaffeination.

Modeling, Simulation, and Optimization of Supercritical and Subcritical Fluid Extraction Processes, First Edition. Zainuddin A. Manan, Gholamreza Zahedi, and Ana Najwa Mustapa.
© 2022 by the American Institute of Chemical Engineers, Inc.
Published 2022 by John Wiley & Sons, Inc.

Knowledge and understanding of the properties of SCF are vital prerequisites for engineers and scientists to capitalize on SFE as a specialized technique for the recovery of valuable components. This chapter presents the fundamental principles and the applications of SFE technology.

1.2 Supercritical Fluid Properties

A SCF is a substance that exists above its critical temperature (T_c) and critical pressure (P_c). Figure 1.1 shows the phase diagram for a pure compound illustrating the supercritical region where an SCF may exist. The fluid above the T_c and P_c cannot be liquefied regardless of the applied pressure and demonstrates unique properties that are different from those of gases or liquids under standard conditions. The fluid also exhibits higher densities that resemble liquids, and lower viscosities that resemble gases. SCF has unique and desirable properties that make it suitable for performing a challenging extraction process. This includes the ability of the fluid properties to change with a slight variation in pressure and temperature near the critical point. High densities for SCFs

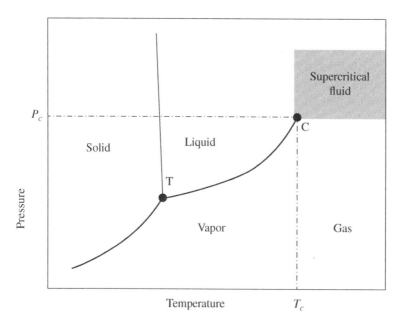

Figure 1.1 A typical P-T diagram for a pure component.

Table 1.1 Comparison of the properties of supercritical CO_2 and those of ordinary gases and liquids.

Substance	Density (g/mL)	Viscosity (g/cm·s)	Diffusion (cm²/s)
Gases	10^{-3}	10^{-4}	10^{-1}
Supercritical CO_2	0.3–1	10^{-3} to 10^{-4}	10^{-4} to 10^{-6}
	Liquidlike	*Gaslike*	*Liquidlike*
Liquids	1	10^{-2}	10^{-5}

contribute to greater solubilization of compounds, while lower viscosities allow SCFs to penetrate solids better and to flow with less friction. Surface tension and heat of vaporization are relatively very low for SCFs (Castro et al. 1994). Since solute diffusivities in SCFs are typically higher than those of liquid solvents by an order of magnitude, and their viscosities are lower also by an order of magnitude, their mass transfer properties are much more favorable.

Table 1.1 shows the properties of supercritical CO_2 compared with typical liquid and gas properties. These unique properties become a major advantage of the SFE particularly for the recovery of thermally sensitive compounds as well as the production of high-purity products. By employing the SFE method, compounds can be extracted at low operating temperature, and highly pure solutes can be easily recovered from the solvent.

1.3 Subcritical Condition

"Near-critical" or "critical region" is defined as the area around the critical point of a solvent. It comprises sub- and supercritical conditions of state for the solvent (Brunner 1994). The critical temperature is the highest temperature at which a gas can be converted to a liquid by the increase of pressure, while the critical pressure is the highest pressure at which a liquid can be converted to a gas by an increase in the liquid temperature. A fluid heated to above the critical temperature and compressed to above the critical pressure is called a SCF. Meanwhile, a pure component is considered to be in a subcritical state if its pressure is higher than the critical pressure and its temperature is lower than the critical temperature. The regions of both fluids are shown in Figure 1.2 as an example of a substance's phase diagram. The subcritical

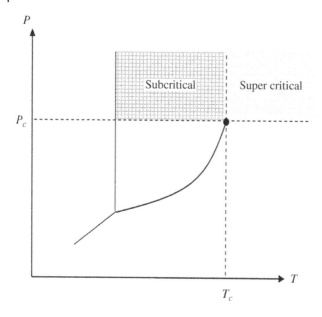

Figure 1.2 An example of phase diagram of pure substance indicating a point of supercritical and subcritical region.

region is located above the critical pressure and below the critical temperature (Brunner 1994). Behavior of phase boundaries on the left of the supercritical region is comparable and does not change dramatically. Moreover, the high-pressure liquid region (subcritical) has many of the characteristics of SCF and is exploited in similar ways (Brunner 1994).

A subcritical fluid is simply a substance that resembles a gas but exists as a compressible fluid that takes the shape of its container. It is liquid but the density and solvating power of the fluid is greater than a conventional liquid since the fluid compressed higher than its critical pressure. Their low viscosity and low-surface tension allow them to penetrate solute matrix and enable rapid wetting. Their higher liquid densities enable them to dissolve solutes from a solid or liquid matrix (Castro et al. 1994)

Brunner (2005) reported that the solubility of analytes in subcritical (liquid) fluid (solvent) increases at constant pressure up to temperatures slightly below the T_c of the solvent. Therefore, it is preferred to operate at or near the critical temperature of the SCF and adjust the pressure in order to obtain optimal fluid density for the extraction to be carried out. In addition, in many cases, the enhancement of the solubility is especially significant near the critical temperature of the solvent (Walas 1985). According to Catchpole and Proells (2001), the use of new near-critical solvents, with the same benefits of CO_2 but substantially lower operating pressures, could increase the applicability of near-critical fluids as solvent. In fact, Illés et al. (2000) in the investigation on extraction of essential oils using SC-CO_2 and propane at super- and subcritical conditions concluded that oil with high biological and commercial value could be produced with compressed gases at super- or subcritical conditions.

Extraction at subcritical conditions has become an increasingly popular alternative method in the extraction of bioactive compounds from natural products, food waste, herbal plants, and in environmental applications. Subcritical water extraction (SWE) is one of the most studied field in the application of environmental analysis, plants, food by-product such as recovered polyphenols from potato peel (Singh and Saldaña 2011), pomegranate (He et al. 2012) algae, and microalgae (Herrero et al. 2006). Subcritical water is also known as superheated water, pressurized hot water, or hot liquid water at temperatures between 100 and 320 °C, and pressure below the critical pressure (220 bar) that maintains water in liquid state. In the extraction of polyphenols, subcritical water offered better extraction yield as compared to extraction using methanol or ethanol. In addition, by using subcritical water, it is possible to extract a different type of free and bound form of phenolic compounds (Singh and Saldaña 2011; Zeković et al. 2014).

1.4 Physical Properties of Subcritical Fluid

As mentioned in the previous section, subcritical fluids have many similar characteristics to SCFs. In fact, subcritical fluid behaviors are exploited in the same manner. Discussions in this book may refer mostly to CO_2 as a supercritical solvent primarily because CO_2 is the most well established and most commonly used solvent in SFE applications. However, the SFE principles also apply to solvents other than CO_2 because the principal behaviors and the characteristics of subcritical and supercritical regions are similar.

The key property that contributes to the sub and SCF characteristics is solvent density. Density of the critical fluids depends on pressure and temperature (Turner et al. 2001). In addition, the density is extremely sensitive to minor changes in temperature and pressure near the critical point. The typical density of both critical fluid changes in a similar way where the density in both regions increases sharply with increasing pressure at a constant temperature, and also decreases with increasing temperature at a constant pressure. The density of SCF is similar to that of a liquid, whereas the subcritical fluid has a higher density since the phase exists as a high-pressure liquid. The solvent power of a fluid increases with density at a given temperature and could increase with temperature at a given density. However, polar solvents exhibit more marked changes in their dissolving power with density increase compared to less-polar solvents (Castro et al. 1994).

The properties of SCFs are frequently expressed in terms of reduced, rather than absolute values (Castro et al. 1994). Figure 1.3 is a phase diagram of both the supercritical and subcritical regions. In the subcritical liquid region, the greatest density changes can occur, and is therefore most effective for changing density when minimal temperature and/or pressure changes are introduced. To achieve subcritical liquid condition, the pressure of a solvent should be applied above the critical pressure, while the temperature is maintained below the critical temperature.

As in the case for density, diffusivity, and viscosity of a sub- or SCF are also related to temperature and pressure. The diffusivity and viscosity of a SCF approach a liquid behavior as pressure is increased. The diffusivity of a solute in a subcritical fluid decreases with pressure increase at a constant temperature and always exceeds the solute diffusivity in an ordinary liquid solvent. On the other hand, viscosity increases with pressure at a given temperature (Castro et al. 1994). In addition, the transport properties of near critical or subcritical fluid are almost similar to those of SCFs. On the other hand, the solvating power of near critical fluid is stronger than that of SCF (see Figure 1.4).

Polarity is one of the key properties that have strong influence on solubility. It can be altered in order to modify the selectivity of an extraction process.

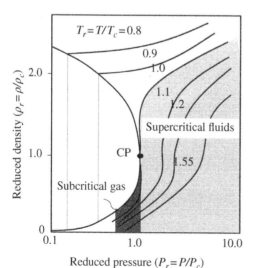

Figure 1.3 Reduced density–reduced pressure including several reduced temperature for supercritical and subcritical region. Source: Castro (1994).

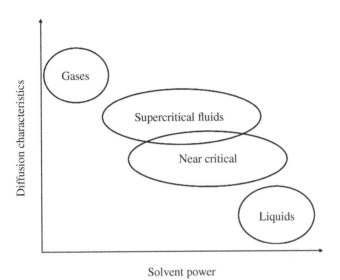

Figure 1.4 Relative solvent power and diffusion characteristics of liquids, near-critical and SCFs as well as gases. Source: Griffith (2001). The reference: Griffith, K.N. (2001). Environmentally Benign Chemical Processing in Near- and Supercritical Fluids and Gases Expanded Liquids. Thesis for Degree Doctor of Philosophy in Chemistry. Georgia Institute Technology.

A pressure rise or the presence of a modifier does not increase the polarity of a fluid as this is determined by the dipole moment of the fluid molecules. CO_2 is a nonpolar solvent, and this limits the solubility of several polar compounds in CO_2. Therefore, the addition of an entrainer is needed to enhance the solubility (not increase the polarity of solvent) of polar compounds. Hansen et al. (2001) has found that polar halocarbons such as 1,1,1,2-tetrafluoroethane should be more able to dissolve polar groups such as capsaicin molecules.

1.5 Principles of Sub- and Supercritical Extraction Process

Subcritical fluid is also known as a high-pressure liquid. Figure 1.5 shows a typical phase diagram for a pure material and the thermodynamic states of various separation processes. Even though the terminology is different from SCF; however, the characteristics and principles of extraction are identical. The basic element required for conducting SFE includes a fluid source, a compressor to pressurize and pump the fluid, an extraction vessel to hold the

1 Fundamentals of Supercritical and Subcritical Fluid Extraction

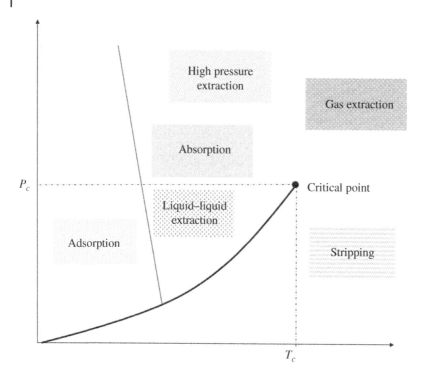

Figure 1.5 Phase diagram of a pure material and the thermodynamic states of various separation processes. Source: Rizvi, S.S.H., Yu, Z. R., Bhaskar, A. R. and Chidambara Raj, C. B. (1994). Fundamental of Processing with Supercritical Fluids. In: Rizvi, S. S. H. (Ed) Supercritical Fluid Processing of Food and Biomaterials, Blackie Academic and Professional. Glasgow NZ, Chapman & Hall, (pp. 1-26).

material to be extracted, a temperature/pressure control system, a collection device, and a backpressure regulator. The backpressure regulator is used to allow a drop in pressure which will enable separation of extract from solvent (Brunner 1994). Additionally, other equipment such as valves, flow meters, and heater/coolers for temperature control of the fluid are needed for proper operation of the process (Rivizi et al. 1986). In industrial applications, the consumption of CO_2 is high, thus a recycle system is needed to control and minimize the cost of the material consumption (Herrero et al. 2006).

1.5.1 Solid Sample Extraction

Extraction of solutes from solids can be represented as a two-stage process comprising extraction and separation of the extract from the solvent (see Figure 1.6). During the extraction step, the solvent is first compressed to above

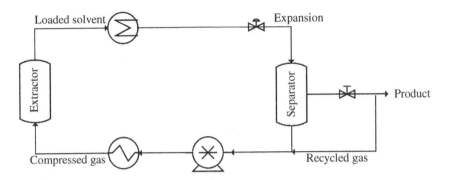

Figure 1.6 Flow diagram of a separation process.

critical pressure before it flows through a fixed bed of solid particles in the extractor and dissolves the extractable components of the solid. The loaded solvent is removed from the extractor and fed to a precipitator. In a large-scale SFE process, the solvent is typically recycled to the separation stage. Solvent losses are compensated by a make-up stream (Brunner 1994).

Collection of the extract yield is a key step in an SFE process. In order to maximize sample collection efficiency, techniques such as cooled collectors and liquid traps should be included. Sample collection can be accomplished by simply allowing the solvent/extract mixture to depressurize completely to atmospheric pressure while the solvent is evaporated and dissipated, leaving only the extracted material in a collection vial. Otherwise, pressure could be reduced to the level which is enough to decrease the solubility of the extract in the solvent. This can be done without complete depressurization in a separation vessel held above atmospheric pressure. In this way, the solvent can be recycled and the energy costs associated with pressurizing the solvent can be saved.

1.5.2 Liquid Sample Extraction

The process equipment for carrying out separation by SCF consists of mainly two columns as shown in Figure 1.7. The first column acts as a separation column for the removal of less-volatile substances, while the other serves as a precipitation column for the separation of the extracted less-volatile components from the SCF cycle.

The mixture to be separated is normally introduced into the middle or on top of the separation column, while the compressed supercritical solvent flows upward through the column; thus, creating a counter current flow. Separation

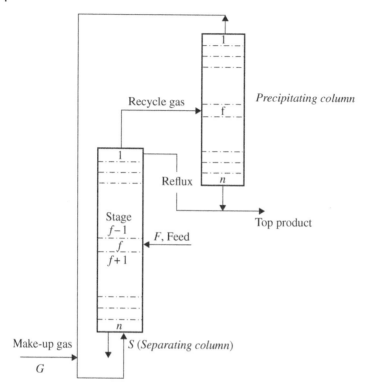

Figure 1.7 Schematic of a SFE process.

of low-volatility components takes place in this column. The component in the mixture which is more soluble in the SCF phase is removed from the mixture by supercritical solvent and becomes enriched in the up-flowing supercritical solvent. Operating conditions of the separation column are at a constant temperature above the critical temperature of the SCF in the range of 1.1–1.5 times the reduced temperature of the SCF; and a pressure of about 2–3 times the reduced pressure of the SCF (Brunner and Peter 1982).

The second column (precipitation column), that is mainly used for solvent recovery, can be operated on different schemes: (i) operates at the same pressure as the separation column except with increased temperature; (ii) operates at the same temperature as the separation column, but the pressure in the precipitation column is reduced. Both techniques drastically reduce the solubility of the low-volatility components in the SCF phase. The liquefied substances flowing downward in the column and are divided into the top product (also known as the extract) and the reflux for the separation column. The regenerated supercritical solvent leaves the second column at the top and is recycled via a circulating pump into the separation column.

The liquid reflux flows downward through the separation column in a counter-current direction to the SCF stream. The reflux changes its composition on account of mass transfer in the column between liquid and SCF. The liquid stream that is withdrawn at the bottom of the separation column is called raffinate. It contains components that are less soluble in the supercritical solvent.

Pressure and temperature are the main operating parameters in controlling an SFE process. Generally, the solubilities of fats and oils are increased by increasing both pressure and temperature (Stahl et al. 1988). Therefore, if a maximum extractability is required, higher operating pressure and temperature are applied. Application of SFE at low pressure and temperature can remove undesirable odors from products such as vegetable oils (Koseoglu et al. 1996).

1.6 Applications of SCF Extraction

Plant materials contain highly valuable compounds such as essential oils, pigments, vitamins, antioxidants, and aroma. Extraction of these compounds is important in the development of natural-based products in food, pharmaceuticals, and nutraceuticals. Conventionally, the compounds are extracted by Soxhlet extraction with the use of organic solvent at their boiling point condition. However, this method may cause products degradation for the thermolabile compounds at high temperatures. In addition, use of organic compounds can lead to product contamination and cause toxicity for human consumption. SFE has been recognized over many years as an alternative green extraction method for a wide range of natural products such as spices, medicinal plants, coffee, and many more.

The first SFE plant was developed for commercialization in the late 1970s. The patented plant was invented by Zosel (1974) for decaffeination of coffee in Bremen, Germany, and operated at 60 000 metric tons per year. Since then, many applications of SFE appeared mainly in industries such as food, pharmaceuticals, nutraceuticals, and polymers.

1.6.1 Decaffeination of Coffee and Tea

Caffeine is a natural compound also known as xantine, a class of alkaloid that naturally occurs in tea and coffee beans, Kola nuts, cocoa, and other few herbs plants. The compound is beneficial for the human body due to its stimulating effect on the central nervous, circular system, and muscle which could increase concentration, energy availability, and have positive effects on Alzheimer's disease. However, excessive consumption of the compound can cause serious health problems to pregnant women, infants, and children. In fact, higher daily doses of caffeine may lead to insomnia, restlessness, and anxiety (Nehlig 1999). The maximum average allowance in daily caffeine intake for an average person is 400 mg/person/day. The allowance is 200 mg/person/day for a pregnant woman and 75 mg/person/day for 10-year-old children. Higher dose of the caffeine consumption may result in miscarriage and heart problems. Therefore, decaffeination of coffee beans or caffeine-based products is an innovative option for human consumers to maintain good health. Besides, the recovered caffeine can be used as additive material for the food and pharmaceutical industry.

The extraction of caffeine from coffee beans can be performed either by traditional technique, i.e. solid/liquid extraction using chlorinated hydrocarbon such as dimethyl chloride, chloroform, ethyl acetate, and ethanol at atmospheric pressure or alternatively by a greener and cleaner extraction method using fluids at critical pressure. Conventionally, coffee beans undergo five processing steps:

1) Pretreatment 1: Crushing process where the beans are grounded to appropriate size.
2) Pretreatment 2: Humidifying the beans with superheated steam to swell it, and to enhance the caffeine extraction from the solid material. The water content may range from 15 to 30% w/w wet basis.
3) Extraction of caffeine by organic solvent such as chloroform, isopropanol, ethyl acetate, and benzene.
4) Recovery of the decaffeinated beans by steaming at vacuum pressure to evaporate the residual organic solvent.
5) Drying at elevated temperature or vacuum until appropriate bean moisture is achieved.

Despite the high selectivity and solubility of caffeine in organic solvents due to its toxicity, the use of the organic solvents for caffeine extraction from coffee beans can pose health hazards to humans. Moreover, the process of organic solvent removal from the coffee beans needs longer time to completely evaporate the solvent. Such prolonged exposure of the beans to high temperatures can adversely affect coffee flavor. SFE provides an effective and safe alternative method for caffeine extraction. The extraction of caffeine was first introduced by Zosel (1974) in his patent on the decaffeination of green coffee beans using moist supercritical carbon dioxide at temperature range of 40–80 °C and pressure from 120 to 180 atm within 5–30 hours of extraction time (Zosel 1981). Since then, numerous patents reported on further development of the decaffeination of coffee from coffee beans or tea.

Caffeine has poor solubility in supercritical CO_2 due to its polar characteristic. The solubility of caffeine in supercritical CO_2 can be enhanced in the presence of other cosolvents such as water. Between 0.5% and 15% w/w of cosolvent can be added, depending on the weight ratio of coffee to supercritical CO_2. The cosolvent can either be mixed with supercritical CO_2 prior to entering an extraction column, or introduced at the entrance of the column, where it is mixed with supercritical CO_2 before it touches the beans.

Figure 1.8 illustrates a cosolvent-assisted extraction process described by Zosel (1982). Supercritical CO_2 solvent is introduced to the extraction column which contains some water at the bottom of the column below a basket containing coffee beans. Prior to extraction, the coffee beans underwent pretreatment processes that include crushing and humidifying to swell the beans. The purpose of the beans swelling is to ease the extraction of caffeine from the

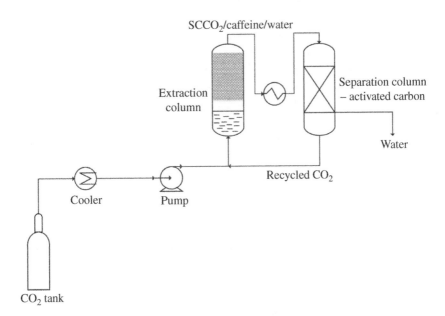

Figure 1.8 Decaffeination of coffee beans by supercritical CO_2 with activated carbon.

coffee beans. It is important to note that excessive use of water in the treatment may cause the coextraction of other water-soluble constituents that could destroy the coffee aroma and flavor. The amount of water must therefore be controlled so that it will only swell the beans without excessively draining caffeine from the material. For an efficient extraction of caffeine from the solid, the water content in the beans should be controlled to between 25 and 50% by wet weight basis (Katz 1989; Zosel 1982).

To begin the extraction process, the column is heated to the extraction temperature of between 40 and 80 °C using the jacket heater on the vessel. CO_2 is then pumped to its supercritical pressure (120–180 bar) and fed through the bottom of the column where it passes through a water reservoir. This causes water–gas bubbling as water is absorbed into CO_2 at supercritical states. Saturated supercritical CO_2 with water fraction of between 1 and 3% has been reported to improve the extraction of caffeine (Katz 1989). At this condition, the supercritical CO_2 becomes humidified and tends to facilitate the extraction of the caffeine due to the increase in solubility of the caffeine to be absorbed into the supercritical CO_2. Saturated stream comprising supercritical CO_2-caffeine-water leaves at the top of the column and enters a depressurization vessel that separates caffeine from the supercritical stream.

As shown in Figure 1.8, to recover caffeine, the supercritical CO_2-caffeine-water stream is cooled down from 70 to 25 °C to liquefy the supercritical CO_2 and condense water before CO_2-caffeine enters the second separation column. The separation column is filled with activated carbon, and the temperature is maintained at 25 °C. In the column, caffeine is absorbed by the activated carbon, and water is collected while the dry supercritical CO_2 is recycled to the extraction column. This process can remove as high as 99.99% caffeine, whereas the residual (0.01%) remaining in the decaffeinated coffee beans may function as aromatic properties. The supercritical CO_2-water decaffeination process may take between 5 and 30 hours.

In other separation techniques, the stream of supercritical CO_2/caffeine/water enters counter-currently with polar solvent in an absorber column to extract the caffeine from the stream, as shown in Figure 1.9. Water is the most preferred polar solvent to extract caffeine from the supercritical CO_2 stream, but may vary depending on solvent polarity. The most important criteria affecting the effectiveness of the caffeine absorption is the optimum weight ratio of the SCFs to the polar solvent. In general, the supercritical CO_2 stream and the polar solvent are contacted at a weight ratio of between 10:1 and 20:1 (Katz 1989). The operating conditions of the absorber closely match the temperature and pressure in the extraction vessel.

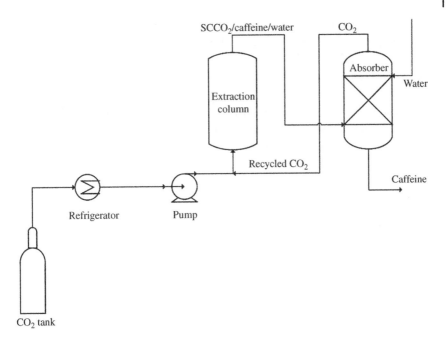

Figure 1.9 Decaffeination of coffee by supercritical CO_2 and absorber.

The separated CO_2 is then recycled to the extraction column. Extraction is carried out for at least eight hours to reduce the caffeine content to about more than half of its initial content in the beans. In the finishing step, the swelled decaffeinated coffee beans are recovered by drying at high temperature, or by vacuum through the flow of air or nitrogen.

1.6.2 Removal of FFA in Fats and Oils

SFE technique has also been applied for oils and fats isolation in industry. Most of the fatty oils are still produced by conventional processes which involve the thermal degradation of products. Hence, the development of new extraction and fractionation processes is necessary. SFE of natural products using CO_2 has already been studied in an attempt to replace conventional processes which require high energy due to the high operating temperature, in the fats and oils industry.

The application of supercritical CO_2 in separating and fractionating fatty acids from natural resources is one of the most promising developments in

SCF technology. Some examples are the refining of palm oil (Ooi et al. 1996), degumming of soybean oil (List et al. 1993), deacidification of rice bran oil (Dunford et al. 2003), and refining of olive oil (Bondioli et al. 1992; Brunetti et al. 1989; Gonçalves et al. 1991; Zahedi et al. 2010c). In addition to that, SFE technique has also been investigated to recover oil fractions with high concentration of vitamins especially carotenes and tocopherols (Brunner et al. 1991; Gast et al. 2001; Oghaki 1989). The following section provides an overview on the type of separations that have been investigated in oil and fat processing using supercritical CO_2.

High free fatty acids (FFA) in vegetables oils can cause rancidity and lower the quality and functionality of the oil (Wei et al. 2012). Meanwhile, in the biodiesel uses, the presence of FFA can encourage soap formation, cause difficulty in separation of product, and lower the yield of biodiesel (Berchmans and Hirata 2008). Therefore, the removal of FFA is very important in order to maintain the best oil quality.

Most of the work related to the separation of fatty acids and triglycerides deals only with the solubility of triglyceride or fatty acid in supercritical CO_2 in the vapor-phase composition. Chrastil (1982) measured the solubility of certain fatty acids and triglycerides in supercritical CO_2 at the pressure range of 8–25 MPa and temperature range of 40–80 °C. Bamberger et al. (1988) measured the solubility of triglycerides and fatty acids in supercritical CO_2. They reported that the solubilities of trilaurin and palmitic acid are about the same, although their molecular weights differ by a factor of about 2.5. They attributed this to the higher polarity of free fatty acids as compared to that of triglycerides. Bharath et al. (1992) used a static apparatus to measure phase equilibrium for oleic acid-CO_2 and triolein-CO_2 systems with the aim of separating the fatty acid from triglycerides and fractionating mixtures of triglycerides. Maheshwari et al. (1992) compared the solubility isotherms of fatty acids and vegetable oils, and suggested that separation of fatty acids from triglycerides might be possible by using supercritical CO_2 at densities below 700 kg/m^3. Brunner (2000) concluded that removal of FFA from the crude oils (triglycerides) is relatively easy and can be completed in a few theoretical stages with particularly no loss in triglycerides based on the separation factor between FFA and triglycerides.

Brunetti et al. (1989) suggested the use of supercritical CO_2 extraction as a method for olive and husk oil deacidification. Gonçalves et al. (1991) carried out the SFE process on the olive oil, and the experimental results suggested that the deacidification of olive oil was feasible. Results from Bitner et al. (1986) suggested that supercritical CO_2 extraction could eliminate caustic refining and bleaching steps in soybean oil refining. List et al. (1993) developed a SCF-based process which permits the countercurrent refining (treatment) of soybean oil to produce a refined quality oil suitable for direct deodorization.

In biodiesel application, SCE oil from *Jatropha curcas* and fractionation was employed simultaneously to reduce the FFA content, maximize the yield, and improve the quality of oil. FFA was effectively removed by fractionation by up to 26.3 wt% in the first fraction at 200 bar. Further increase in pressure to between 300 and 500 bar in the second and third fractionation stages resulted in good oil removal performance with low FFA content (1.43 wt%) and yielded oil as high as 47 wt% oil yield. SFE offered a cost-effective processing alternative technology that has the potential to eliminate the degumming, esterification, and dehydration steps (Fernández et al. 2015).

1.6.3 Enrichment of Tocopherols

Use of supercritical CO_2 to concentrate tocopherols from the distillate of a deodorizer was initiated in the 1980s. The main challenge of the separation process is in separating tocopherols and FFA from vegetable oils, and from deodorizer condensates comprising essentially of FFA, monoglycerides (MG) and diglycerides (DG), tocopherols, and sterol compounds that are produced during refining of crude oils. Processes for extracting tocopherols from edible oils using CO_2 were patented by Ohgaki and Katayama (1992) and Uno et al. (1985).

The feasibility of supercritical CO_2 as a promising technique for concentrating natural tocopherols (>50%) with recovery as high as 80% from the by-product of soybean oil refining has been demonstrated (Fang et al. 2007; Mendes et al. 2005; Shi et al. 2011). Economic evaluation has shown that the process can be profitable if the plant produces above 900 tonne/year of deodorizer distillate for soybean oil (corresponding to 25% of rate of production) (Mendes et al. 2002). Apart from by-product of soybean oil, the use of supercritical CO_2 to enrich tocopherols has been employed to herbal plants and fruits such as olive tree leaves (De Lucas et al. 2002), grapes seed (Bravi et al. 2007), Kalahari melon and roselle seeds (Nyam et al. 2010), amaranth (Kraujalis and Venskutonis 2013), *Aloe Vera*, and almond (Bashipour and Ghoreishi 2014). In general, pressure increase has significantly increased tocopherols content due to the increase in solvating power. To get a better solubilization of tocopherols, addition of a modifier, i.e. 5% of ethanol into SC-CO_2 gave higher yield than when pure CO_2 was used (Kraujalis and Venskutonis 2013). Nevertheless, there was no clear effect found on the influence of temperature on the extraction of tocopherols (De Lucas et al. 2002).

1.6.4 Carotenes from Crude Palm Oil and from Palm Fatty Acid Esters

Ooi et al. (1996) concluded that it is not possible to extract carotenes from triglycerides using SFE even with the addition of the entrainers due to the low solubility of carotene in CO_2 and low concentration of carotene in Crude Palm Oil (CPO). In terms of selectivity, that means that all other components can be readily separated as top product from the CPO. However, carotenes can only be separated as a bottom product. This implies that 99% or more of the feed have to be removed as a top product. Considering the relatively lower solubility of triglycerides in supercritical CO_2, the exhaustive extraction of palm oil can therefore be economically prohibitive.

Another possibility is to modify the CPO to esters. Gast et al. (2001) showed that separation of esters from carotenes can easily produce a highly concentrated mixture of carotenes which can be treated effectively using countercurrent SFE. Fatty acid methyl esters or fatty acid ethyl esters produced from the transesterification of CPO is used as biodiesel fuel.

1.7 Solubility of Solutes in SCFs

The effectiveness of SFE processes is demonstrated by its ability to extract compounds from materials. This is characterized by the solubility of solutes in SCFs. The solubility of solutes is defined as the equilibrium concentration, or an amount (e.g. gram of solute extracted from a material by 1 g or 1 mol of supercritical solvent. A very high solubility of a compound in SCFs indicates that the compound can be effectively separated by SFE technique. On the other hand, SFE method is less-favorable for a compound with a very low solubility in a SCF. Reliable data obtained through SFE experimental works is essential for the design of a full-scale SFE system and for the success of SFE operation.

A supercritical solvent increases the solubility of the solute significantly near its critical region. This phenomenon can be noted from the relationship between the partial molar volume of the solute in the SCF phase and the derivative of the fugacity coefficient with respect to pressure. In Eq. 1.1, it can be noted that the partial molar volume of the solute (v_2) in the SCF phase increases significantly due to small changes in pressure (P) and/or temperature (T) near the critical region because the pressure derivative of fugacity coefficient is large at this condition.

$$\left(\frac{\partial}{\partial P} \ln \varphi_2\right)_{T,x} = \frac{v_2}{RT} - \frac{1}{P} \tag{1.1}$$

Figure 1.10 Variation of the reduced density of CO_2 in the vicinity of critical point. Source: McHugh and Krukonis (1994).

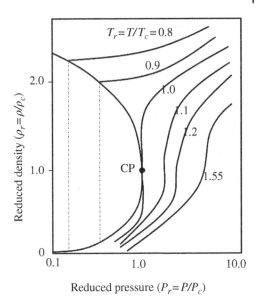

Figure 1.10 demonstrates the changes in the density of CO_2 with pressure at different temperatures. Around the critical pressure (P_c), a drastic change in the density of the solvent occurs with a small change in pressure. Since the solubility of solutes in SCF increases with an increase in solvent density, a dramatic change in solubility is expected by an increase in pressure from a level below the P_c (i.e. subcritical pressure) to a level above P_c (i.e. supercritical pressure). It is also shown in Figure 1.10 that solvent density decreases with an increase in temperature in the supercritical region.

Solubility of solutes in supercritical solvent relies mainly on thermodynamic factor especially temperature and pressure. Solvent density exhibits the solvating power that can be altered by changing pressure, temperature, or both. Increase in solvent density, which is as a function of both temperature and pressure, increases the solubility of solutes. These are the key physicochemical properties that are essential for SFE.

Other factors that influence the solubility of compounds are polarity of solvent, the solute properties including its volatility (or vapor pressure), molecular weight, and polarity. These properties influence the solute–solvent interactions and are important in obtaining a detailed understanding of supercritical extraction potential (Bruno 2006).

1.8 Solute–Solvent Compatibility

Chemical and physical properties of solute and solvent are important in influencing the solubility data. Based on the "like-dissolves-like" principle, polar solute will dissolve in polar solvent, whereas nonpolar solute will dissolve in nonpolar solvents. Vast differences in the degree of polarity between solute and solvent will lessen the solubility of solute in the particular solvent. For example, the presence of polar groups in the solute structure decreases its solubility in the nonpolar solvents. This chemical property is associated with the dielectric constant and dipole moment of the solvent and explained its effect on solubility (Corr 2002).

In terms of solute properties, the solubility of solutes is also influenced by vapor pressure and molecular weight. Low vapor pressure and high molecular weight tend to limit the solubility of a solute in a solvent as compared to highly volatile solutes with low molecular weight. For a lower molecular weight solute, its solubility is dependent on the functional group present. On the other hand, the solubility of a higher molecular weight solute is dominated by its polarity as shown in Figure 1.11. As an example, triglycerides with short chain and low polarity fatty acids are more soluble and can be easily removed from the SC-CO_2 extracting solvent. This explains why the solubility of low volatility substances in nonpolar supercritical gases decreases with increasing

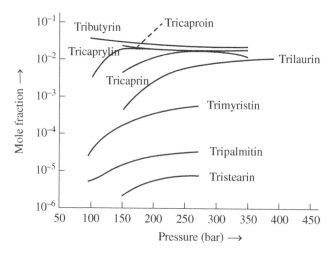

Figure 1.11 Solubility of components of homologous series in SC-CO_2 (*Source:* Adapted from Brunner (1994). © John Wiley & Sons).

molecular mass and with increasing polarity and number of polar functional groups (Brunner 1994).

On the other hand, the higher the solutes vapor pressure, the easier the removal of the solute from the matrix binding. Therefore, as the solute vapor pressure increases the solubility increases (King and Catchpole 1993). Moreover, a polar and high-molecular weight analyte requires a modifier to raise its solvating power (Castro et al. 1994). Thus, solvent modifiers such as ethanol or methanol are essential to make the analyte and solvent polarity compatible.

In terms of solute volatility factor, all in all, the most important factor affecting the solubility is the process operating conditions, namely, pressure and temperature.

1.9 Solubility and Selectivity of Low-Volatility Organic Compounds in SCFs

In general, SFE exploits the large solvent power of dense gases. SFE process has an advantage over liquid extraction due to the fact that SCF exhibits gas-like viscosity and diffusivity while retaining liquid-like densities (McHugh and Krukonis 1994). The solvent power is directly related to the thermodynamic state of the solvent and may be controlled by manipulating temperature and pressure. In understanding the advantages of SFE, it is important to know the supercritical solvent properties and their interactions with the organic components to be separated. The properties of a thermodynamic system consisting of a supercritical solvent and substances of low volatility can be summarized as follows:

1) The solubility of low-volatility organic components in a SCF depends mainly on the density of the supercritical solvent. As temperature increases, at constant pressure, the density of the SCF decreases and likewise the solvent capacity of the solvent for low-volatility components decreases. On the other hand, with increasing temperature, the vapor pressure of components to be dissolved also increases. According to the dominant effect, the solubility increases with temperature, and vice versa. Nonetheless, the difference in solvent capacity cannot be explained solely by density effects.

2) Gases, which have similar critical temperatures, dissolve low-volatility substances in different quantities. As an example, the solubility of a mixture of glycerides (i.e. palm oil, consisting up to about 90% of triglycerides) in

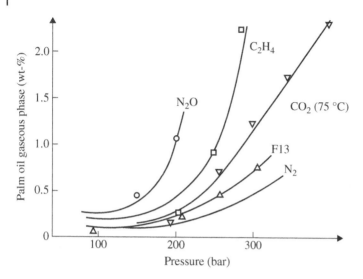

Figure 1.12 Solubility of palm oil in various SCFs (*Source:* Brunner and Peter (1982). © Elsevier).

various SCFs is plotted in Figure 1.12. The results show that at given temperatures and pressures, the solubility of substances in SCFs increases with the critical temperature of supercritical solvent. However, it was found that gases like carbon dioxide (CO_2), nitrous oxide (N_2O), and chlorotrifluoromethane (CF_3Cl), all of which have a critical temperature of about 30 °C, dissolve different amounts of glycerides. Therefore, specific interactions must therefore be taken into account.

3) The different solvating capacities of supercritical solvents are also related to the solubility of the supercritical solvent in the liquid phase. Figure 1.13 shows the solubility of different SCFs in palm oil. Note from the figure that, the higher the solubility of a solvent in palm oil, the better the oil is dissolved by the solvent. The effects of changes in temperature and pressure on the solubility of a substance in a SCF depend on the influence of these changes on the equilibrium concentration of the supercritical solvent in the liquid phase.

4) Solvent capacity of a given supercritical solvent increases exponentially with pressure. In most cases, fairly high pressures have to be applied in order to obtain significant concentrations in the SCF phase. The solubility of palm oil in SCFs at 70 °C and at pressures of up to 150 bar is considerably

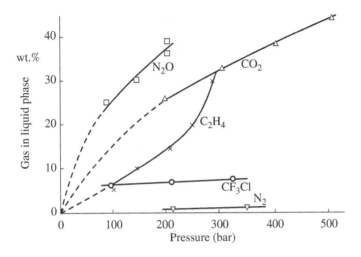

Figure 1.13 Solubility of various SCFs in palm oil (*Source:* Brunner and Peter (1982). © Elsevier).

below 0.5 wt%. Above 200 bar, the palm oil concentration in the SCF phase increases rapidly (see Figure 1.13).
5) Chemically similar substances dissolve in SCFs according to their volatility (vapor pressure). More volatile components dissolve in SCF to a greater extent. The same applies to chemically different substances, strongly differing in volatility. In the case of substances, which differ in association, the solubility in the SCF phase is not proportional to their differences in volatility (i.e. water/ethanol in supercritical CO_2).
6) The solubility of components to be separated in the SCF phase can be enhanced in the presence of an entrainer with volatility lying between those of the components to be separated, and the supercritical component. Figure 1.14 shows how separation factors can be influenced by specifically adding entrainers. Compared to a pure supercritical component as an extracting agent, the operation can be carried out at lower pressures in the presence of an entrainer. Besides that, the temperature dependence of concentration of low-volatility substances in the supercritical phase is much more pronounced than in the absence of an entrainer.
7) SCF can selectively dissolve certain components from mixtures whose components have similar volatility, but different chemical structure. Experimental determination of selectivity (K-factors) of components in SCF is essential. Figure 1.14 shows the selectivity of FFA of palm oil plotted

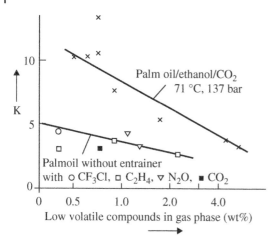

Figure 1.14 Selectivity (K-factors) of FFA in palm oil (*Source:* Brunner and Peter (1982). © Elsevier).

against the amount of low-volatility matter dissolved in the supercritical phase. The higher the K-factor, the more FFA dissolves in the SCF phase. The lesser line represents a mean value for various supercritical solvents without an entrainer. The upper line represents the K-factors when ethanol is added to CO_2 as an entrainer. In this case, the K-factor is roughly doubled.

1.10 Method of Solubility Measurement

Basically, solubility data must be obtained experimentally by several reliable methods such as dynamic, static, or a combination of both. Several techniques have been developed to measure the solubility of a solute in SCFs. Methods of solubility measurement fall into four categories, namely the dynamic (flow) method, static (or equilibrium) method, chromatographic method, and spectroscopic method (Bruno 2006). The static and dynamic methods are, however, the main methods that are most frequently used for solubility measurement.

1.10.1 Static Method

In static or equilibrium methods, a fixed amount of solute and solvent is loaded into a high-pressure cell which may include a sight window. Fluid recirculation is continuous and agitation is applied to reach equilibrium rapidly.

Temperature and pressure in the cell are adjusted to dissolve all or a portion of the solute in SCF. Samples of the fluid phases can be taken and analyzed as necessary. A binary system sampling may not be needed and conditions may be adjusted to reach the equilibrium by visual observation through the windowed cell. This method is often used to eliminate the need to sample the SCF solution. However, most of the data reported for CO_2 in literature are obtained in flow-through apparatus using the dynamic method. The static method is however preferred when oil present in a solid sample is in low concentration. The dynamic extraction method is superior to the static method when the solid sample is high of oil concentration (Castro et al. 1994).

1.10.2 Dynamic Method

Dynamic or flow method for solubility measurement is the simplest and most straightforward technique as summarized by Bruno (2006). Using this technique, fresh liquid solvent at ambient temperature is charged and compressed with a high-pressure pump to the desired pressure. The pump is used to deliver fluid continuously passed through a section of tubing or a preheating column to ensure that it is at the desired temperature and then passed over through the sample matrix inside the extraction vessel which thermal equilibrium with operating temperature in constant-temperature bath or oven (Taylor 1996).

The fluid then flows through a high-pressure vessel containing the solute and the fluid needs to be saturated with solute. The fluid then leaves the pressure vessel and depressurizes to atmospheric conditions using a heated metering valve or restrictor. At the same time, the extracted solute will precipitate or fall out of the fluid solution, collected and quantified gravimetrically, or by using some conventional analytical technique. The solvent volume can either be measured using a totalizer or by measuring the flowrate of the solvent and the collected sample (Taylor 1996).

When using the dynamic technique, it is important to ensure that the solute and fluid solvent can reach equilibrium when the solvent floods the extractor column. Therefore, measurements at a number of solvent flowrates need to be done since the equilibrium conditions will be determined at the appropriate solvent flowrate. Bruno (2006) stated that if the calculated solubility is independent of flowrate, equilibrium is usually assumed. However, the main disadvantage of the dynamic flow method is that it requires the test to ensure the equilibrium condition to be performed for every system at every temperature studied. Other disadvantages of this technique are given by Bruno (2006), McHugh and Krukonis (1994), and Ashraf-Khorassani et al. (1997).

Table 1.2 summarizes the advantages and disadvantages of the dynamic method.

After obtaining the data experimentally, the solubility data can be calculated from the initial slope of the extraction curve: oil yield (g/g) *versus* solvent consumption (L/g sample), as shown in Figure 1.15 and in Eq. 1.2. The initial data of the curve can be fitted by least square approximation method or by using a

Table 1.2 The advantages and disadvantages of solubility measurement using the dynamic method.

Advantages	Disadvantages
Simple and fast	Bed and flowrate channeling problem might happen in extractor
Large amount of solubility data obtained reproducibly and rapidly	Solvent–solutes concentration could be below equilibrium solubility if poor mass transfer between solvent–solutes
Fractionation and equilibrium data can be obtained	Plugging of solid or liquid sample in restrictor
Straight forward sampling procedure	Decomposition of thermally-sensitive compounds compounds through a heated metering valve or restrictor

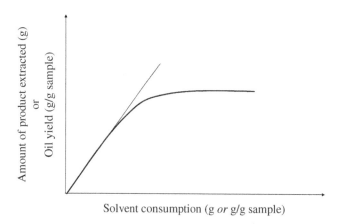

Figure 1.15 Solubility calculation by method of dynamic experimental.

program such as curve fitting. The slope of the linear curve is therefore known as experimental solubility.

$$\text{Solubility}(g/g) = \frac{\text{mass of extracted oil (g/g sample)}}{\text{volume of solvent consumption (L/g sample)}}$$

(1.2)

1.11 Determination of Solvent

Selection of the suitable solvent for the extraction is important to ensure high extraction yield and selectivity of the products. The primary factor to select a suitable solvent for a separation process is its ability to extract the desired product from a mixture. Therefore, the compatibility between solutes and solvent is also an important factor to be considered in order to perform a reliable solubility measurement. Detailed discussion concerning the factor will be presented in the next section. According to Rivizi et al. (1986), the solvent choice should have a low boiling point for ease of removal after processing, and finally it should be inexpensive, nontoxic, nonflammable, and readily available in high purity. Table 1.3 summarizes the specification that should be considered when choosing an appropriate solvent for an SFE process. The ideal properties as listed in the table are important in order to achieve the best performance or the maximum separation yield.

Other key factors influencing solvent selection are critical temperature and pressure. An ideal solvent should have a low or moderate critical pressure to minimize compression cost (Rivizi et al. 1986). Basically, high critical temperature may destroy heat-sensitive solutes. The temperature range commonly employed in supercritical extraction falls between 30 and 150 °C. Therefore, in order to avoid a thermal degradation or deterioration for heat-sensitive solutes, the solvent must have critical temperatures lying in this range (Taylor 1996). Table 1.4 shows the critical properties of various fluids commonly used in SFE technology.

Water has a very high critical temperature. Ammonia is toxic, whereas hydrocarbon group such as propane is flammable and forms explosive mixtures with the air. Even though Sulphur Hexafluoride (SF6) has the lowest critical pressure, its solvating power is insufficient for all aromatic and polar analytes. Carbon dioxide has been well accepted as the most employed solvent in many SCF extraction due to its advantages such as inert, nontoxic, and nonflammable. In addition, CO_2 has a low critical temperature of 31.1 °C which

1 Fundamentals of Supercritical and Subcritical Fluid Extraction

Table 1.3 Advantages and disadvantages of the most commonly used SCFs.

Properties	Inorganic				Organic		
	CO_2	NH_3	H_2O	N_2O	CFCs	HC	MeOH
Polarity	—	✓	✓	✓	✓	—	✓
Toxicity	✓	—	—	—	—	—	✓
Inflammability	✓	—	✓	✓	✓	—	—
Reactivity	✓	—	—	—	✓	✓	—
Cost	✓	—	✓	—	✓	✓	—
Ease of reaching supercritical conditions	✓	—	—	✓	✓	✓	—
Ease of recovery	✓	✓	—	✓	✓	✓	✓
Environmentally aggressiveness	✓	—	✓	—	—	—	—
Gaseous under ambient conditions	✓	✓	—	✓	✓	✓	—

CFCs: chlorofluorocarbons; HC: hydrocarbons; MeOH: methanol.

Table 1.4 Various solvent as SFs with critical properties.

Solvent	T_c (°C)	P_c (bar)
Sulfur hexafluoride	45.5	37.1
Chlorotrifluoromethane	29.0	38.7
1,1,1,2-tetrafluoroethane	101.2	40.6
Propane	96.8	42.4
Trichlorofluoromethane	198.0	43.5
Trifluoromethane	25.9	46.9
Chlorodifluoromethane	96.4	48.5
Xenon	16.6	57.6
Dinitrogen monoxide	36.5	71.7
Carbon dioxide	31.0	72.9
Ammonia	132.4	111.3
Water	374.0	217.7

can help to prevent thermal degradation of thermally labile food components. Despite, several researchers also have pointed out its disadvantages mainly polarity and fairly higher critical pressure than others solvents. Thus, it typically needs a higher operating pressure to perform a successful extraction. Consequently, a much higher capital and operating costs are needed.

Several works have reported on the uses of Freon gases mainly R134a and R12 and other gases, for example, nitrous oxide as the fluid had slightly higher polarity in contrast to CO_2 and suitable for polar compounds. Nevertheless, R134a was more promising in terms of lower critical pressure, polarity, and has been recognized as nonozone-depleting since it does not contain chlorine molecules (Catchpole and Proells 2001; Simoes and Catchpole 2002).

In chemical properties perspective, Castro et al. (1994) explained that nonpolar and scarcely polar solvents which have moderate critical temperature (e.g. N_2O, CO_2, ethane, propane, pentane, xenon, SF6, and some Freon) have a limited dissolving power for solutes of a high polarity or molecular weight. Thus, small proportions of modifiers are added to enhance their dissolving power by enhancing the solubility of polar compounds into CO_2 which is a highly nonpolar solvent. However, the addition of polar entertainers such as methanol could increase the effective critical temperature, and this can damage thermally labile products. In addition, solvent modifiers may also need a downstream separation to obtain a pure product. In contrast, R134a is a naturally polar solvent and therefore the solubility of polar compounds in R134a is rather sufficient without the addition of any modifier. R134a also has a relatively lower critical pressure compared to other solvents. Therefore, this can minimize the operating cost due to lower pressure required during the SFE process.

Other relevant physical properties to consider for a SCF as a solvent are viscosity and surface tension (see Table 1.5). R134a has both low viscosity and low surface tension; allows rapid wetting, and penetration of solvent into its sample matrix structure. Its low viscosity is also encouraging rapid solute diffusion through the solvent, thus promoting good rates of solute dissolution (Corr 2002). Table 1.5 shows that R134a has a lower viscosity and surface tension compared to most conventional liquid solvents and is comparable to SC-CO_2.

Carbon dioxide is the common supercritical solvent used in food applications. It is not only cheap and readily available at high purity but also safe to be handled, and it is easily removed from food by expansion to ambient pressure. The practical advantages of CO_2 are also applicable to R134a. Carbon dioxide becomes supercritical at a critical temperature (T_c) of 31.1 °C, whereas R134a will change to its supercritical condition at 101.4 °C. Even though the T_c

Table 1.5 Physical properties of some common solvents (Corr 2002).

Solvent	Density (g/cm^3)	Viscosity (mPa·s)	Surface tension (dyn/cm)
SC-CO$_2$ (200 bar, 40 °C)	0.84	0.1	NA
R-134a	1.21	0.21	8.7
Diethyl ether	0.71	0.223	17.0
Ammonia	0.77	0.138	—
Hexane	0.66	0.294	18.4
Acetone	0.79	0.316	23.7
Chloroform	1.49	0.41	—
Ethanol	0.79	1.04	22.4
Methanol	0.79	0.55	22.6
Water	1	0.89	73.0

At 25 °C.

of R134a is much higher than that of CO$_2$, experimental conditions near the critical temperature are considered sufficient as the compressibility is reasonably good in order to obtain a variation of solubility data (Hansen et al. 2001). Therefore, R134a seems to be a particularly attractive solvent for the food industry since it meets all the aforementioned criteria.

1.11.1 Carbon Dioxide (CO$_2$)

CO$_2$ is the most commonly employed SCF solvent due to it being nontoxic, nonflammable, inert, and inexpensive. It is present as a SCF above its critical temperature and pressure of 31 °C and 73.8 bar. CO$_2$ has been technically approved by Federal Drugs Administration (FDA) as odorless, colorless, and safe for food and pharmaceuticals-based applications. CO$_2$ critical temperature that is near to the ambient temperature makes it suitable for extraction, reaction, impregnation, drying, and emulsion processes that involve sensitive-heat compounds such as polyphenols, essential oils, and organic compounds. Its moderately low critical pressure makes it feasible and widely applied in industry for applications such as decaffeination of coffee and tea, extraction

Table 1.6 Dielectric constants and dipole moments for various solvents.

Solvent	Dielectric constant, ε (kHz)	Dipole moment, DM
R-134a	9.5	2.05
Hexane	1.9	0.08
Diethyl ether	4.34	1.52
Dichloromethane	9.08	1.55
Tetrahydrofuran	7.61	1.63
Acetone	20.7	2.9
Methanol	32.6	1.66
Acetonitrile	36.6	3.92
Dimethyl sulfoxide (DMSO)	47.2	3.96
Water	80.0	1.85

of hops and tea, recovery of spices, aroma, vitamin E, phytosterol, and several fatty acid methyl esters.

Table 1.6 shows the properties of supercritical CO_2 in comparison to typical liquid and gas properties. Solvating power of supercritical CO_2 is associated with its density, while the density is strongly dependent on the temperature and pressure. Therefore, the solvation strength can simply be tuned by adjusting the process temperature and pressure to selectively extract desired solutes. On the other hand, manipulation of the temperature and pressure also would change the diffusivity and viscosity of the CO_2 becoming a gas-like property, hence, increasing diffusion and mass transfer of the solvent into pores or intact cells of solid materials. These unique properties become a major advantage of the SFE particularly for the recovery of thermally sensitive compounds as well as the production of high-purity products. By employing the SFE method, compounds can be extracted at low operating temperature, and highly pure solutes can be easily recovered from the solvent.

1.11.2 1,1,1,2-Tetrafluoroethane (R134a) as a Solvent

R134a is the replacement refrigerant for (chlorofluorocarbons – CFCs) since the conventional refrigerant R12 (CFC 12) has been phased out due to its harmful effect on the environment. R134a is a hydroflourocarbons (HFCs) substance containing only atoms of hydrogen, fluorin, and carbon and are known to have effectively zero-ozone depleting potential (ODP) at the time the Montreal Protocol was established (Corr 2002). R134a is used in refrigeration, polymer foam blowing, as an aerosol propellant and as a solvent for special cleaning applications. R-134a is available at high purity, thus it is used as a medical propellant for inhalation applications such as metered dose inhaler (MDI) (Barker et al. 1999).

R134a has also been approved for food use in Europe in compliance with the terms of the FDA. A wide range of R134a extracts is approved in Japan. This includes extraction of vanilla from Madagascar Bourbon beans and extraction of fragrance such as jasmine flower. The extracts of R134a are generally low in color, low in inert lipid content, and high in the desired impact or active species with good recovery of the active ingredients from the raw materials. Besides, R134a is also regarded as Generally Recognised as Safe (GRAS) affirmation for food-grade extracts in the United States (Corr 2002).

However, R134a is a much more expensive fluid than CO_2 and has a global warming potential (GWP) of 1300 times higher than that for CO_2 (Barker et al. 1999; Wood and Cooper 2003). Therefore, a recycling system of R134a needs to be considered for economic and environmental reasons. R134a systems design needed to be handled with care to minimize loss of R134a during the process (Simoes and Catchpole 2002). Energy-efficient recycling of R134a may be practical since it was developed originally as a refrigerant (Wood and Cooper 2003). However, the impact of R134a on climate change is only minor in comparison to the total impact of CO_2 emissions. This is because CO_2 as a greenhouse gas arises mostly from the burning of fossil fuels and mainly accumulates in the atmosphere (McCulloch 1999).

R134a or HFC 134a is classified as a halocarbon family refrigerant. Its chemical name is 1,1,1,2-tetrafluoroethane (CF3CH2F). It is manufactured from the reaction of hydrogen fluoride with trichloroethylene in a closed system (Barker et al. 1999). Figure 1.16 shows the molecular structure of R134a. R134a (1,1,1,2-tetrafluoroethane) is nontoxic, nonreactive, nonflammable, and nonozone-depleting (McCulloch 1999; Wood et al. 2002). It also has a high volatility and a boiling point of −26.2 °C at atmospheric pressure, which means that it leaves negligible solvent residues in a product. R134a is a nonflammable halogenated fluid, stable-to-aqueous acids and bases and can be regarded as an aprotic, which is in general, a

Figure 1.16 Molecular structure of 1,1,1,2-tetrafluoroethane.

1.11 Determination of Solvent

chemical term. It is immiscible with water and sparingly soluble in water (1500 ppm at 20 °C). It is normally handled as a compressed gas under pressure in liquid form and has liquid density of about 1.3 kg/L (Corr 2002).

R134a has a similar ability compared to CO_2 in terms of product recovery. Separation of R134a from the extracted product can be conducted efficiently at relatively low temperature while achieving minimal solvent residues in the isolated product (Corr 2002). This is due to the high volatility of R134a.

Generally, the solvent strength or solvent power can be easily controlled because it depends on temperature and pressure. In recent years, the solvent properties of R134a have been under examination. The focus has been on the properties of supercritical R134a (Abbott and Eardley 1998), supercritical mixture of CO_2, and R134a (Abbott et al. 1999) and vapor–liquid equilibrium of the binary mixture of carbon dioxide-R134a (Duran-Valencia et al. 2002). In mixtures of CO_2 and R134a, R134a has been found to be an effective cosolvent or modifier, acting to increase the solubility of polar solutes and to increase the eluotropic strength of the solvent mixture over that of pure CO_2 (Abbott et al. 1999).

Roth (1996) concluded that, in the high pressure and low-compressibility region, the solvating power of the six refrigerants (HFC: R32, R125, R134a, and R152a, HCFC: R22 and R123) are greater than that of carbon dioxide. Specifically, the solvent power of R134a can be readily tailored to the desired level of extraction or separation by adjusting the temperature and pressure of the supercritical phase.

The solvating powers of R134a actually are influenced by two factors: the density and polarity of solvent. The polarity function is usually defined as dipole moment (μ) and dielectric constant (ε). The μ for R134a is 2.05 while for CO_2 is 0.00. Since R134a is more polar than supercritical carbon dioxide, which is regarded as nonpolar over a wide range of operating pressure, R134a is more able to form a strong intermolecular interaction with polar solutes within the liquid phase of R134a. Therefore, R134a would result in better solubility than supercritical CO_2. Table 1.6 shows some values for dielectric constant and dipole moment for a number of solvents including R134a. Both parameters are indicators of solvent polarity. It is obvious that the polarity of R134a is higher than several conventional solvents (Corr 2002). The solvents with higher polarity constant such as acetone and methanol are not suitable for SFE application due to their toxicity and flammability (Castro et al. 1994; Corr 2002).

The solubility of species in a SCF is governed by the relationship between the solvent density and polarity/polarizability parameter, $\pi*$. The effective $\pi*$ of a solvent depends on the density of SCFs. Increasing the density will increase the $\pi*$, and hence, the solubility. Abbott and Eardley (1998) found that the $\pi*$ value in the liquid region of R134a increases roughly linearly with increasing pressure in the liquid state and the value of $\pi*$ decreases dramatically at temperature above 100 °C and below 60 bar. Therefore, R134a shows a large change in solvent properties over a relatively small temperature and pressure range, suggesting that it would be a useful solvent for extraction purposes.

The use of R134a as a solvent in supercritical extraction was first invented by Blackwell et al. (1996). They who studied the solubility of polynuclear aromatic hydrocarbons found that R134a in a supercritical state is suitable for use in supercritical extraction. However, due to high critical temperature of R134a ($T_c = 101.9$ °C), Ashraf-Khorassani et al. (1997) on their trial investigated the solubility of drugs in subcritical R134a and claimed that the solubility was significantly higher than the solubility in supercritical CO_2 and chloroform. These results have opened for numerous research works on the application of R134a in the extraction, polymer, and thermodynamic area.

In the United Kingdom, Dean et al. (1998) has reported the first extraction of natural products using R134a for the extraction of magnolol from Magnolia officinalis. The extraction process was named as "Phytonics" that utilized liquefied R134a as a solvent which was known as "Phytosol A." Nevertheless, the yield obtained with pure R134a was reported as the lowest compared to the yield by supercritical CO_2 + methanol. As for natural product application, R134a can be used to extract a range of materials (Wilde 2001). Products that may be extracted include natural flavors and fragrances and nutraceutical extracts. The extracts obtained using R134a are generally low in color, low in inert lipid content, and high in the desired impact or active species with good recovery of the active ingredients from the raw materials.

In addition, there were numerous researchers who also studied R134a on extraction and solubility as well as fractionation of active compounds. These include studies on the solubility of β-carotene from palm oil (Mustapa et al. 2011), capsaicin and β-carotene (Hansen et al. 2001), extraction of fixed and mineral oils (Wilde 2000), solubility of lipids (Catchpole and Proells 2001), extraction of terpene lactones and flavonoids from Ginkgo leaves (Chiu et al. 2002), and solubility of vanillins (Knez et al. 2007).

Wilde patented processes of extraction of fixed (e.g. palm oil) and mineral oils (petroleum) using liquefied R134a year (2000) and iodotrifluoromethane year (2001) at the pressure range 20–25 bar and temperature 40–60 °C with or without cosolvent. It has been found that R134a can be used to extract a range of useful products from a wide range of materials of natural origins. Wilde

(2000) has surprisingly found that R134a, though a very poor solvent for fixed (e.g. palm oil) and mineral oils (e.g. petroleum) at low temperatures, is actually a very much better solvent at elevated temperatures. At 40 °C for example, cocoa butter (a fixed oil which comprises natural mixtures of mono-, di-, tri-glycerides, fatty acids, sterols [and their esters] and natural waxes) dissolves in R134a to a substantial extent. Therefore, at a temperature only a few degrees lower which at room temperature cocoa butter does not dissolve to any appreciable extent in R134a. In addition, the fat is some solutes such as fatty acids and triglycerides are only slightly soluble even in hot R134a at temperatures between 40 and 60 °C in particular of liquid extraction.

R134a also has been suggested as an alternative to the use of CO_2 in supercritical extraction. Higher polarity and lower critical pressure are the main reasons R134a appeared as the best alternative to CO_2. In the extraction of carotenoids and chlorophyll, a from Laminaria japonica Aresch (Lu et al. 2014) and astaxanthin from *Euphausia Pasific* (Han et al. 2012) has demonstrated that subcritical R134a is a potential solvent to supercritical CO_2 under mild condition. On the other hand, Catchpole and Proells (2001) who measured the solubility of lipids in subcritical R134a in a countercurrent packed column at low pressure ranging from 40 to 200 bar at 70 °C claimed that the solubility of oleic acid in R134a below 120 bar and 70 °C is appreciable even at pressures as low as 45 bar as compared to the acid's relatively very small solubility in CO_2 at the same conditions. They have shown in their prior work that the solubility of all lipids in supercritical CO_2 at 60 bar and temperatures of 40–80 °C is effectively zero. The subcritical R143a was concluded to be a viable and cost-effective alternative to supercritical CO_2 since much lower operating pressure could be used to extract higher percentages of lipids. Simoes and Catchpole (2002) who evaluated the potential of using subcritical R134a for extracting valuable substance from lipid mixture also arrived at the same conclusion. The raffinate contents of squalene are almost identical for the supercritical CO_2 and subcritical R134a for the extract product. In the same work, they found that the separation of squalene from triglycerides/glycerol ethers fraction using subcritical R134a was feasible.

In comparison to several compressed fluorinated hydrocarbons (R23 and R236fa) as well as supercritical CO_2, Knez et al. (2007) determined equilibrium solubilities of vanillins, ethylvanillin, *o*-vanillin, *o*-ethylvanillin in those gases with static-analytic method in a pressure range between 110 and 260 bar and temperature from 40 to 60 °C. The highest solubility of vanillin in R134a was reached at 203 bar and 60 °C and that is approximately 3.8 times higher than in CO_2 at the same pressure and temperature. The solubility *o*-vanillin in R134a is in the same order of magnitude and is comparable as in CO_2. Both vanillins with ethoxy group are less soluble in R134a than in CO_2 (ethylvanillin up to approximately 2 times, o-ethylvanillin up to 12 times).

Beyond the application of natural product, Wood et al. (2002) reported on the use of R134a as an alternative to supercritical CO_2 in polymerization applications and considered the solvent to be a promising substitute to SC-CO_2 in terms of environmental and economic aspects. Besides, subcritical R134a also has been applied in the removal of cholestrol from spray-dried *Sthenoteuthis oualaniensis* egg powder (Sui et al. 2014). The removal efficiency using subcritical R134a was found significantly higher than those achieved by using supercritical CO_2 with the rate as high as 99.16%.

1.12 Important Parameter Affecting Supercritical Extraction Process

For a successful application of SFE technology, several critical factors or parameters must be taken into consideration in conducting SFE experiments. These factors include the type of sample, method of sample, type of fluid, and extraction conditions: pressure, temperature, solvent flowrate, and extraction time (Lang and Wai 2001). According to Reverchon (1997), those parameters could significantly affect the extraction rate. However, the main factor that affects the extraction performance is the operating conditions, namely pressure and temperature. In addition, the extraction performance is also indirectly affected by solvent flowrate, cosolvent, moisture content, solvent-to-feed ratio, and raw materials.

1.12.1 Pressure and Temperature

Extraction pressure is the most relevant parameter to describe the behavior of SFE processing technology. In general, increasing pressure of fluid at constant temperature increases the solvent density and enhances the solvent power. The solvent power is described in terms of solvent density at the given operating conditions. Reverchon and De Marco (2006) states that higher pressure increases the solvent power and reduces the extraction selectivity. Turner et al. (2001) also noted that the fluid density should not be higher than necessary in order to obtain quantitative recovery of the analytes since higher fluid densities generally would lower the analytes selectivity. For example, Birtigh et al. (1995) has demonstrated that higher pressure results in poorer selectivity for the extraction of tocopherols and carotenoids from palm oil components due to increasing solubility of triacylglycerols.

Temperature is another key parameter that contributes to the successful SFE of fat-soluble vitamins which require careful optimization of the extraction temperature (Turner et al. 2001). In the context of SFE processing, an increase in fluid temperature may increase, decrease, or have no effect on the solubility of the solute depending on the pressure. The solutes volatility and diffusivity tend to increase with increasing temperature. The fluid density, however, may decrease with temperature at a constant pressure.

Indeed, the temperature effect is more complex than the pressure effect as the former influences the solvent density and the solutes volatility, and ultimately, the extraction rates. Temperature increase (at a fixed pressure) reduces the solvent density, thereby reducing the solvent's solvating power. Despite, the solubility still can increase due to the increases of the solutes' vapor pressure when the temperature increased and improved the tendency of the solutes to dissolve in the fluid phase (Reverchon and De Marco 2006). This phenomenon is called as "crossover" due to competing effect between decreasing of solvent density and increasing solutes vapor pressure as temperature increase (de Melo et al. 2009; Duba and Fiori 2015; King and Bott 1995; Özkal et al. 2005) can be seen from a plot of solubility against pressure at different temperatures (Figure 1.17) (Salgın and Salgın 2013). Below and above the

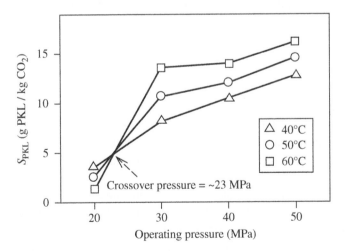

Figure 1.17 Crossover intersection in plot of solubility versus pressure at different temperature (*Source:* Salgın and Salgın (2013). © Elsevier).

intersection is known as lower and upper crossover pressure, respectively. The increase of solubility with pressure (at constant temperature) above the crossover point is due to the increase of solvent density and hence its solvent power. However, increases of temperature at constant pressure is more pronounced and results in higher solubility at high pressure and temperature. On the other hand, below the crossover point, the solubility is dominantly influenced by the density of solvent. The increase of temperature has little effect to decrease the density of solvent and to increase solute vapor pressure. Thus, higher solubility is obtained at lower temperature and pressure.

Note that temperature has a significant influence on the desorption kinetics and may determine the extraction rate of adsorbed solutes (Castro et al. 1994) and resulted in higher oil yield toward the end of extraction, as has been observed by Duba and Fiori (2015). In addition, higher amounts of extract also can be attributed to the solute's volatility and/or mass transfer rates increasing with temperature (Brunner 1994).

1.12.2 Solvent Flowrate

Solvent flowrate is another important parameter to consider in solubility measurement. If the flowrate is too high, the solvent–solute saturation could not be attained. On the other hand, when flowrate is low enough used, it is believed that the saturated of solute–solvent could be achieved. There are some ranges of flowrate that would give roughly similar solubility value. Larger flowrates may decrease the solubility value due to reduced residence time, which ultimately affects the solvent–solute equilibrium. Higher solvent flowrate may result in the extractor to be simply bypassed without sufficient solvent–solute (Saldaña et al. 2002). Besides, King (1997) stressed that some adjustment is required for each individual matrix when conducting SFE. He also noted that when extracting seeds with high oil content, lower solvent flowrate through the bed was found to be advantageous in order to avoid compaction of the spent meal that may prevent complete extraction of the oil.

Even though solubility is primarily dependent only on pressure and temperature which affects solvent density and solute vapor pressure; however, solvent flowrate is considered as indirect factors contributing toward oil solubility. This is because, in order to attain equilibrium condition, it is desired to provide a sufficient residence time for solvent–solute to achieve saturation. Therefore, low flowrate may prolong the achievement of solvent–solute equilibrium, whereas high flowrate may completely prevent equilibrium from being achieved since further flowrate increase may cause the solubility to drop.

Failure to reach equilibrium is one of the important sources of error when using the dynamic method. Therefore, a few key factors were considered during the selection of optimum flowrate. These include solubility, oil yield, extraction time, and solvent consumption.

An approach to determine whether an extraction process is controlled by external mass transfer or intraparticle diffusion resistance involves constructing the curve of % yield (g oil/g sample) or mass extracted (g) against solvent consumption (g) at different flowrates. This plot can determine whether the extraction process either dependent or independent of flowrate. If all the extraction curves at different flowrates overlap, this indicates that the extraction process is independent of flowrate and is controlled by external mass transfer or solubility. When this happens, solvent–solute equilibrium is considered to exist at the outlet of the extractor (Sovová 2005, 2012). This phenomenon has been reported in studies done by Tonthubthimthong et al. (2001), Kiriamiti et al. (2002), and Salgın et al. (2006). If, on the other hand, intraparticle diffusion resistance predominates, extraction rate will depend on the flowrate. This can be observed when higher yields are obtained at lower flowrates and vice versa (Goto et al. 1993). Such behavior was reported by Papamichail et al. (2000).

1.12.3 Cosolvent

Supercritical CO_2 is known as a nonpolar solvent and has low solubility for extraction of polar solutes. To improve the extractability of the polar compounds, cosolvent is employed as a modifier to the supercritical CO_2 in an extraction system. Addition of cosolvent to the supercritical CO_2 may affect the phase behavior of the extraction process due to the formation of charge transfer complex between solutes and solvent (Ekart et al. 1993). In general, the use of cosolvent may improve the solvent power of supercritical CO_2 and increase the selectivity of the compounds toward the supercritical CO_2. A significant improvement of the extraction yield was observed from 0.68 to 3.07% yield/dry weight material at 40 °C and 400 bar by supercritical CO_2 + ethanol as cosolvent (Pimentel et al. 2013). The yield of extraction from leaves of *Piper piscatorum* was reported to show less variation in comparison to the yield obtained with the supercritical CO_2 + methanol. The increase of extraction yield improved by relying on the polarity of the solvent used. Studies showed that addition of water resulted in a substantially higher extract yield in comparison to ethanol, methanol, and dimethyl sulfoxide (Casas et al. 2007). Greater polarity of the cosolvent might result in higher extraction of the polar solutes.

Ethanol and water are the most commonly organic solvent used as a modifier for the extraction of bioactive polar compounds from solid materials due to their nontoxic and safe for edible products and pharmaceutical application. For the extraction of lipids, ethanol is more favored, while for the extraction of glucosinolate and phenolic compounds, water has been found as more efficient at higher pressure and temperature (Solana et al. 2014). Nevertheless, in some cases, a mixture of ethanol-water resulted as the most promising modifier for the extraction of high valuable compounds. For example, use of 10% of ethanol–water mixture (57% v/v) as cosolvent has been proven yielding a rich fraction of phenolic compounds at 8 MPa and 40 °C with 6 kg/h of CO_2 flowrate. In addition, it is important to note that when the flowrate of supercritical CO_2 is constant, a system with lower concentration of ethanol–water mixture provides most remarkable extraction of phenolic compounds with substantial high antioxidant activity (Da Porto et al. 2014). In the extraction of *Tabernaemontana catharinensis* extracts from its leaves, the average global yield was found constant as $2.4 \pm 0.1\%$ when pure alcohol cosolvent was used and significant improved as high as $15 \pm 1\%$ when a mixture of alcohol–water is employed in the supercritical CO_2 extraction (Pereira et al. 2005).

On the other hand, ideal percent of the cosolvent is important to take into account in the optimization studies. Excess percent of cosolvent utilized in the supercritical CO_2 extraction could lead to strong interaction between cosolvent and solutes through solute/cosolvent bond and reduce the interaction between CO_2 and solutes which then cause the decreasing of the extraction yield (Michielin et al. 2009). In their study, 5% w/w was found as the maximum concentration of cosolvent can be applied in the extraction of *Cordia verbenacea* extracts to obtain higher yield and the yield tend to decreased when 8% w/w of the cosolvent used.

1.12.4 Moisture Content

High moisture content in samples is generally a disadvantage for SFE methods. However, in some cases, the water content is useful to facilitate the oil removal during the extraction process. The water content could act as cosolvent and enhance the solubility of oil in supercritical CO_2. A substantially high amount of water encourages the extraction of polar compounds such as astaxanthin from crawfish shells (Charest et al. 2001) and an essential oil from herb material (Leeke et al. 2002). They reported that the increase of yield was due to the high solubility of oxygenated compounds such as thymol, linalool, and α-terpineol in water which facilitates the mass transfer from herb matrix into supercritical CO_2 by minimizing the intraparticle resistance. A similar finding was reported for the polar aromatic compounds such as monoterpenes and sesquiterpenes. Increase of moisture content modifies the solvent power of supercritical CO_2; hence, increasing its selectivity and affinity toward high polarity and low molecular weight compounds (Ivanovic et al. 2011). They demonstrated that the increase of 40% of extraction rate with the increased moisture content was not due to the coextraction of water, nonetheless, good dissolution of water that acts as an entrainer enhances solubility of solutes. In addition, improved internal mass transfer resistance of solutes owing to the swelling effect of plant materials with the presence of water had increased the extraction of solutes.

Several studies had demonstrated that certain ranges of moisture content had negligible effect on extractability. For example, Snyder et al. (1984) described a sample with moisture content in the range from 3 to 12 wt% had little effect on the solute's extractability. Franca et al. (1999) noted that the moisture content of Buriti (*Mauritia flexuosa*) fruit used ranged from 7.5 to 11 wt% for SCE of lipids and carotene. The ground hazelnut particles had 3% by weight of moisture content (Özkal et al. 2005) prior to supercritical CO_2 extraction. Norhuda (2005) used a palm kernel sample with moisture content between 3 and 4% for extraction using SC-CO_2. Bhattacharjee et al. (2007) first cracked, dehulled, and flaked the cottonseed prior to SFE so that the surface area increases and ruptures a large number of cell walls, resulting in good oil recovery. The moisture content of the cotton seeds was also reduced to around 7% for successful extraction. Azizi (2007) prepared *Pithecellobium Jiringan* seed samples with 4.27% by weight of moisture content.

Excess moisture content may affect extraction process efficiency. Nagy and Simándi (2008) found that moisture content in paprika samples ranging from 7 to 18% showed little effect on the extraction yield. However, if the water content was higher than 18%, the extraction efficiency reduced gradually and extremely low of yield obtained if fresh solid materials with extremely high moisture content, i.e. ~85% was used. The high-water content in samples would cause a barrier for the CO_2 to dissolve oil due to limitation to diffuse into solids matrix. For instance, extraction of phytonutrient, i.e. carotenoids was reduced due to the presence of water that had increased the diffusion resistance of the compounds to dissolve in supercritical CO_2 (Kostrzewa et al. 2020; Sun and Temelli 2006). In addition, high moisture content also led to the coextraction of water with oil and compounds from solid samples, which later requires additional steps to remove water from the extract. This situation happens when solubility of water and oil become similar at specified pressure and temperature.

All in all, it is important to determine the feed material moisture content, and to which level removal of water would be needed prior to an extraction process in order to achieve excellent efficiency. In addition, the information on the initial moisture content is also useful to minimize coextraction of water into supercritical CO_2 that consequently could lead to a more complex post-extraction process. Optimum moisture content has to be measured and a suitable water content in solid samples for a SFE process is dependent on the nature of the solid matrix and types of solutes.

1.12.5 Raw Material

Extraction of solutes using SFE method can be carried out in either solid or liquid state. An important criteria for a successful extraction of solutes from solid material is the physical structure or nature of the solid matrix. Various types of solid samples are usually employed in an SFE method, for examples leaves, root, seeds, flowers, stems, and fruits. These solid materials contain different percentages of initial oil content depending on the nature of the plant. In principle, solutes are trapped in the matrix and possess solute–solid interaction that will affect the mass transfer and extraction rates. To facilitate diffusion of supercritical solvent into the pores of the matrix and hence increase extractability of the solutes, a suitable pretreatment of the solid matrix is required. However, the pretreatment procedures are subject to the initial oil content and type of the matrix.

Fruits and fiber matrices such as avocado, apricot, coconut, palm kernel, and palm mesocarp are usually rich in oil with more than 40–60% of total dry weight material. These oils are abundantly located at the exterior of the plant wall cell; thus, the extraction of solutes is controlled by external mass transfer and could lead to high extraction rate. Therefore, the extraction yield from this type of material can reach nearly to 75 wt%, considering good solubility of the oil in supercritical solvent with optimized solvent flowrate, particle size, and moisture content.

On the contrary, spices and herbs (commonly leaves and seeds materials) such as ginger, orange peel, black pepper, coriander, and neem are containing essential oils. Essential oil is considered as a fraction of plant oil that consists mainly of monoterpenes and sesquiterpenes, oxygenated derivatives compounds such as ketone, aldehyde, and alcohol (Fornari et al. 2012). These compounds are attributed to natural flavor, color, and aroma. In the course of extraction, compounds in essential oil are distributed at the intracellular of plant wall cells. This explained that the kinetics of extraction of essential principles is controlled by diffusion mass transfer. In general, the average essential oil content in plants is within the range of 1–5% of the total weight of dry feed material. In many cases, supercritical extraction yield of essential oils can be obtained with 3–25 wt%, depending on the nature of the solid matrices and the distribution of compounds in the plant matrix.

Both solid and liquid materials have different SFE operation mechanisms that influence the extraction rate differently. For the extraction of solutes from solid material, particles size of sample and moisture content are important factors that affect extraction rate and yield. The extraction kinetics can be explained and described using operating variables, i.e. porosity, initial oil content, height, and diameter of extraction bed and solvent flowrate (Sovová 2005). These parameters are fitted using mathematical models to produce a reliable estimation data which explain the extraction profiles of the samples.

Meanwhile, for liquid samples such as fats and lipids, essential oils, vegetables oils, and mineral oils are operated via fractionation that separates various components in the oils. The extraction process is typically carried out in a packed column in continuous counter current flow to the direction of supercritical solvent. The aim of this configuration is to increase the contact and mass transfer between solvent and liquid sample as well as to prevent carry-over of liquid sample with supercritical solvent. Information on the feed compositions is essential to allow mass and energy balance calculations to be employed in order to determine the feasibility of the separation process.

1.13 Profile of Extraction Curves

The oil extraction profile from crushed natural products or seeds can be divided into two stages according to the type of oils contained in solid particles. The oils include "free oil" and "tied oil" and are located at the outer layer and in the inner of the particle, respectively (França et al. 1999). In order to simplify the evaluation of the experimental extraction curve, the curve can be distinguished into three distinct regions, i.e. constant extraction rate (CER), falling rate extraction period (FER), or transition to diffusion-controlled and diffusion-controlled extraction period (Povh et al. 2001). The concept associated with the regions of the overall extraction curve has been highlighted by Franca and Meireles (1997), França et al. (1999), França and Meireles (2000), and Quispe-Condori et al. (2005).

The plot of overall extraction curves is a plot of the cumulative amount of oil extracted or cumulative oil yield (in g or in g oil/g sample, respectively) against the amount of solvent used (in g) or extraction time (in minute or hour). The first stage describes the extraction of "free oil" from the external surface of particles at constant rate. During this stage, convective mass transfer between "free oil" and solvent phase occurred and was determined by external mass transfer mechanism. This fast-physical process is represented by a straight line on the extraction curve and is known as the solubility-controlled phase or equilibrium solute–solvent. The solubility values are calculated from the slope of the linear part of the overall extraction curve.

In the second stage, when the "free oil" is depleted from the external surface of the particles, the "tied oil" is extracted at a slower rate than during the first stage. This decreased the rate of extraction rapidly. The oils are diffused into the bulk of the solvent through the particles outer layer and controlled by intraparticle diffusion resistance. This part of the mechanism is named as the diffusion-controlled phase. With further increase in extraction time, the extraction curve approaches an asymptotic oil yield indicating that the oil has been depleted from the materials. Figure 1.18 illustrates the three-step extraction course comprising a constant extraction rate (CER), falling extraction rate (FER), and diffusion regions.

From the extraction curve, a mathematical model can be developed to describe the kinetics and physical behavior of the solute dissolution from the material into the solvent bulk phase. Basically, the empirical model of SFE for solid material can be approached by three methods, i.e. heat and mass transfer analogy and differential mass balance integration. However, a general

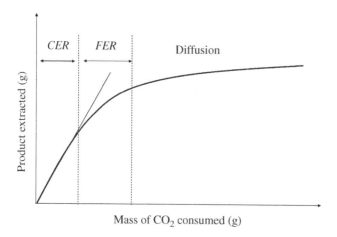

Figure 1.18 Extraction curves indicates the profile of overall extraction curves.

modeling that is usually employed by numerous researchers is based on differential mass balance integration with the assumption is that the mass transfer resistance was only in the solvent phase (Bulley et al. 1984; Lee et al. 1986). Other models that frequently used and reported are known as shrinking core (Ajchariyapagorn et al. 2009; King and Catchpole 1993) and "broken and intact cell" (BIC) model (Döker et al. 2010; Fiori et al. 2009; Machmudah et al. 2006). The aim of each model is to determine the extraction behavior and mass transfer mechanism of the solutes from the internal cells into the solvent. By doing this, the characterization of an industrial scale-based SFE process can be predicted and designed through the simulation of an overall extraction curve.

1.14 Design and Scale Up

The ultimate progress on the application of supercritical extraction method involves the scaling up of the SFE setup, from laboratory scale (50–2000 mL) to pilot scale (5–10 L) extraction vessel (Khaw et al. 2017) and finally, to industrial scale which usually involves an extraction column of more than 100 L. Designing and scaling up of new supercritical extraction process involving solid materials require several important fundamental data. These primarily include thermodynamic equilibrium and process parameters (solubility, temperature, and pressure), kinetics parameters, bed geometry data, and mathematical models that explain the behavior of SFE extraction efficiently.

Scaling up of an SFE process involves several criteria, where these parameters are kept constant: (i) solvent-to-feed mass ratio, (ii) solvent flowrate-to-raw material bed ratio, (iii) solvent-to-feed mass ratio and solvent flowrate-to-raw material bed ratio, (iv) constant of ratio solvent-to-feed mass ratio and solvent flowrate-to-raw material bed ratio and dimensionless ratio, such as Reynolds number, and (v) bed geometrical, i.e. bed diameter and height (Khaw et al. 2017; López-Padilla et al. 2017). In principle, these criteria can be achieved either by developing new empirical equation based on bed geometry or by using kinetics model such as Broken and Intact Cells (BIC) model developed by Sovová (1994), Hong et al. (1990), or Cocero and Garcıa (2001).

The scale up techniques can be done with the aid of kinetics mathematical models to predict and regenerate the overall extraction curve (OEC) with new dimensions of height and diameter of the extraction bed. In this case, solvent flowrate is the most essential factor to be manipulated since it determines the extraction behavior of the SFE; whether it is controlled by external or internal mass transfer. In addition, other critical parameters that contribute to the evaluation of the mass transfer of solvent and solutes within the solid matrix include porosity, particle size, density of extraction bed, and solubility of solutes. Optimization of these parameters is key to the successful design and scale up of a new SFE process.

2

Modeling and Optimization Concept

2.1 SFE Modeling

Mathematical modeling and simulation are the most convenient and efficient way to illustrate the supercritical fluid extraction process compared to its real application. Studying the dynamics behavior of the system and optimization of process parameters are made possible via simulation and hence the economic viability of the SFE process could be estimated.

Mathematical models are usually used to study the effects of parameters and the scaling of the operations to determine their effects on the overall extraction curves (OECs). The overall mathematical model requires construction of two subparts, namely the model of phase equilibrium data for the solute–solvent system and the mass transfer model for the mass transfer behavior in the packed bed. In some detailed models, heat balance and momentum balance equations also may be needed, but generally in most cases, mass transfer model is sufficient to model the process with the help of auxiliary equations and thermodynamic models. The following steps describe general approaches for the modeling:

1) Exact understanding of the system
2) Defining dependent and independent variables of the system
3) Assumptions which will help simplifying and building the model
4) Selecting the system boundaries to apply mass, energy, and momentum balance equations
5) Defining system geometry (such as cartesian, cylindrical, and spherical)
6) Selecting conservation rules for the system (mass, heat, momentum balance, and in some cases, charge balance)

Modeling, Simulation, and Optimization of Supercritical and Subcritical Fluid Extraction Processes,
First Edition. Zainuddin A. Manan, Gholamreza Zahedi, and Ana Najwa Mustapa.
© 2022 by the American Institute of Chemical Engineers, Inc.
Published 2022 by John Wiley & Sons, Inc.

7) Simplifying conservation equations based on step 3 assumptions
8) Defining boundary conditions based on our understanding of the system in step 1
9) Rearranging the equations and using auxiliary equations if needed
10) Selecting suitable numerical solution for solving the equations
11) Analyzing the results and possibly validating the results

Mathematical models have great importance in research and development. In order to develop reaction kinetics and estimating kinetic parameters, scale up of pilot plants, investigating effect of operational condition on plant performance, process optimization, process arrangement and scheduling, investigating interaction of different processes on each other, process control, simulating the process specially during start up and shut down, process troubleshooting, training the staff, and many more information can be obtained from the modeling.

2.1.1 Importance of Knowing the Solid Matrix and Selecting a Suitable Model

A good model for supercritical extraction is a model which is able to predict phase equilibrium, solubility, adsorption, and desorption. Mathematical modeling of SFE is complex because of the complex features of the solid matrices and specially plants. There is not enough knowledge and data bank of the plant matrices. In this case, predicting the plant component behavior during the extraction process is sometimes difficult. This is the reason why most of the studies on SFE modeling of herbs and plants start with understanding plant property. Distribution of solute inside a solid matrix during the extraction process is very important. The distribution in the solid texture plays an important role in model development. In this case, the first step in SFE modeling is the understanding of the solid matrix. The fact has been mentioned in step 1 of the modeling.

2.1.2 Different Modeling Approaches in SFE

As mentioned earlier, mathematical modeling of the SFE from solid matrices becomes difficult because of lack of the information about solute distribution in solid phase and interaction of solute with different components in solid particles. During the years, some models have been developed for modeling the SFE phenomena. These models can be divided in three main categories:

2.1.2.1 Experimental Models
These modeling is based on operational parameters such as temperature, pressure, solubility, particle size, and solute physical properties. When there is not enough data on the mass transfer and equilibrium data these types of models can be used (Papamichail et al. 2000; Subra et al. 1997).

2.1.2.2 Models Which Are Based on Similarity between Heat and Mass Transfer
In these models based on heat and mass transfer analogy, the SFE phenomena are considered similar to heat transfer phenomena. Solid particles are considered to have spherical shape and equations, which are used to describe cooling of hot spheres are employed to describe mass transfer from these spherical solids. In developing these models, all particles are assumed to have the same extraction condition (Reverchon 1996). This is ideal condition which causes error in estimation made these types of the models.

2.1.2.3 Models Based on Conservation Balance Equations
In this type of modeling, mass balance equations are written to model the system behavior. These models include internal and external mass transfers. Usually, mass transfer coefficient is obtained by assuring equilibrium between solid and fluid phase.

2.2 First Principle Modeling

White box, first principle model (FPM) or deterministic model is used when complete information of the process is accessible and the whole governing equations of the system are solvable using analytical or numerical techniques. The governing equations are mass, energy, and momentum balances and empirical equations from thermodynamics and kinetics. The commercial simulators for simulation of oil refining units use FPMs (Worden et al. 2007; Zahedi et al. 2005). Usually, in FPM complex partial differential equations or complex algebraic equations appear which should be solved analytically or numerically.

In order to write the FPM, mass balance known as equation of continuity and momentum balance, known as equation of motion need to be obtained (Bird 1924).

2.2.1 The Equation of Continuity[a]

$$\left[\frac{\partial \rho}{\partial t} + (\nabla \cdot \rho v) = 0\right] \tag{2.1}$$

Cartesian coordinate (x,y,z):

$$\frac{\partial \rho}{\partial t} + \frac{\partial}{\partial x}(\rho v_x) + \frac{\partial}{\partial y}(\rho v_y) + \frac{\partial}{\partial z}(\rho v_z) = 0 \tag{2.2}$$

Cylindrical coordinate (r,θ,z):

$$\frac{\partial \rho}{\partial t} + \frac{1}{r}\frac{\partial}{\partial r}(\rho r v_r) + \frac{1}{r}\frac{\partial}{\partial \theta}(\rho v_\theta) + \frac{\partial}{\partial z}(\rho v_z) = 0 \tag{2.3}$$

Spherical coordinate (r,θ,∅):

$$\frac{\partial \rho}{\partial t} + \frac{1}{r^2}\frac{\partial}{\partial r}(\rho r^2 v_r) + \frac{1}{r\theta}\frac{\partial}{\partial \theta}(\rho v_\theta \sin\emptyset \sin\theta) + \frac{1}{r\sin\emptyset \sin\theta}\frac{\partial}{\partial}(v_\emptyset) = 0 \tag{2.4}$$

When the fluid is assumed to have constant mass density ρ, the equation simplifies to

$$(\Delta \cdot v) = 0$$

2.2.2 The Equation of Motion in Terms of τ

$$\left[\frac{\rho Dv}{Dt} = -\nabla p - [\nabla \cdot \tau] + \rho g\right] \tag{2.5}$$

Cartesian coordinate (x,y,z)[a]:

$$\rho\left(\frac{\partial v_x}{\partial t} + v_x\frac{\partial v_x}{\partial x} + v_y\frac{\partial v_x}{\partial y} + v_z\frac{\partial v_x}{\partial z}\right) = -\frac{\partial p}{\partial x} - \left[\frac{\partial}{\partial x}\tau_{xx} + \frac{\partial}{\partial y}\tau_{yx} + \frac{\partial}{\partial z}\tau_{zx}\right] + \rho g_x \tag{2.6}$$

$$\rho\left(\frac{\partial v_y}{\partial t} + v_x\frac{\partial v_y}{\partial x} + v_y\frac{\partial v_y}{\partial y} + v_z\frac{\partial v_y}{\partial z}\right) = -\frac{\partial p}{\partial y} - \left[\frac{\partial}{\partial x}\tau_{xy} + \frac{\partial}{\partial y}\tau_{yy} + \frac{\partial}{\partial z}\tau_{zy}\right] + \rho g_y \tag{2.7}$$

$$\rho\left(\frac{\partial v_z}{\partial t} + v_x\frac{\partial v_z}{\partial x} + v_y\frac{\partial v_z}{\partial y} + v_z\frac{\partial v_z}{\partial z}\right) = -\frac{\partial p}{\partial z} - \left[\frac{\partial}{\partial x}\tau_{xz} + \frac{\partial}{\partial y}\tau_{yz} + \frac{\partial}{\partial z}\tau_{zz}\right] + \rho g_z \tag{2.8}$$

[a] These equations have been written without making the assumption that τ is symmetric. This means, for example, that when the usual assumption is made that the stress tensor is symmetric, τ_{xy} and τ_{yx} may be interchanged.

Cylindrical coordinate (r,θ,z)[b]:

$$\rho\left(\frac{\partial v_r}{\partial t} + v_r\frac{\partial v_r}{\partial r} + \frac{v_\theta}{r}\frac{\partial v_r}{\partial \theta} + v_z\frac{\partial v_r}{\partial z} - \frac{v_\theta^2}{r}\right)$$

$$= -\frac{\partial p}{\partial r} - \left[\frac{1}{r}\frac{\partial}{\partial r}(r\tau_{rr}) + \frac{1}{r}\frac{\partial}{\partial \theta}\tau_{\theta r} + \frac{\partial}{\partial z}\tau_{zr} - \frac{\tau_{\theta\theta}}{r}\right] + \rho g_r \quad (2.9)$$

$$\rho\left(\frac{\partial v_\theta}{\partial t} + v_r\frac{\partial v_\theta}{\partial r} + \frac{v_\theta}{r}\frac{\partial v_\theta}{\partial \theta} + v_z\frac{\partial v_\theta}{\partial z} - \frac{v_r v_\theta}{r}\right)$$

$$= -\frac{1}{r}\frac{\partial p}{\partial \theta} - \left[\frac{1}{r^2}\frac{\partial}{\partial r}(r^2\tau_{r\theta}) + \frac{1}{r}\frac{\partial}{\partial \theta}\tau_{\theta\theta} + \frac{\partial}{\partial z}\tau_{z\theta} - \frac{\tau_{\theta r} - \tau_{r\theta}}{r}\right] + \rho g_\theta \quad (2.10)$$

$$\rho\left(\frac{\partial v_z}{\partial t} + v_r\frac{\partial v_z}{\partial r} + \frac{v_\theta}{r}\frac{\partial v_z}{\partial \theta} + v_z\frac{\partial v_z}{\partial z}\right)$$

$$= -\frac{\partial p}{\partial z} - \left[\frac{1}{r}\frac{\partial}{\partial r}(r\tau_{rz}) + \frac{1}{r}\frac{\partial}{\partial \theta}\tau_{\theta z} + \frac{\partial}{\partial z}\tau_{zz}\right] + \rho g_z \quad (2.11)$$

[b] These equations have been written without making the assumption that τ is symmetric. This means, for example, that when the usual assumption is made that the stress tensor is symmetric, $\tau_{r\theta} - \tau_{\theta r} = 0$.

Spherical coordinate $(r,\theta,\emptyset)$:[c]

$$\rho\left(\frac{\partial v_r}{\partial t} + v_r\frac{\partial v_r}{\partial r} + \frac{v_\theta}{r}\frac{\partial v_r}{\partial \theta} + \frac{v_\emptyset}{r\sin\emptyset\sin\theta}\frac{\partial v_r}{\partial \emptyset} - \frac{v_\theta^2 + v_\emptyset^2}{r}\right)$$

$$= -\frac{\partial p}{\partial r} - \left[\frac{1}{r^2}\frac{\partial}{\partial r}(r^2\tau_{rr}) + \frac{1}{r\sin\emptyset\sin\theta}\frac{\partial}{\partial \theta}(\tau_{\theta r}\sin\emptyset\sin\theta)\right.$$

$$\left. + \frac{1}{r\sin\emptyset\sin\theta}\frac{\partial}{\partial \emptyset}\tau_{\emptyset r} - \frac{\tau_{\theta\theta} - \tau_{\emptyset\emptyset}}{r}\right] + \rho g_r \quad (2.12)$$

$$\rho\left(\frac{\partial v_\theta}{\partial t} + v_r\frac{\partial v_\theta}{\partial r} + \frac{v_\theta}{r}\frac{\partial v_\theta}{\partial \theta} + \frac{v_\emptyset}{r\sin\emptyset\sin\theta}\frac{\partial v_\theta}{\partial \emptyset} + \frac{v_r v_\theta - v_\emptyset^2\cot\emptyset\cot\theta}{r}\right)$$

$$= -\frac{1}{r}\frac{\partial p}{\partial \theta} - \left[\frac{1}{r^3}\frac{\partial}{\partial r}(r^3\tau_{r\theta}) + \frac{1}{r\sin\emptyset\sin\theta}\frac{\partial}{\partial \theta}(\tau_{\theta\theta}\sin\emptyset\sin\theta)\right.$$

$$\left. + \frac{1}{r\sin\emptyset\sin\theta}\frac{\partial}{\partial \emptyset}\tau_{\emptyset\theta} + \frac{(\tau_{\theta r} - \tau_{r\theta}) - \tau_{\emptyset\emptyset} - \cot\emptyset\cot\theta}{r}\right] + \rho g_\theta \quad (2.13)$$

$$\rho\left(\frac{\partial v_\emptyset}{\partial t} + v_r\frac{\partial v_\emptyset}{\partial r} + \frac{v_\theta}{r}\frac{\partial v_\emptyset}{\partial \theta} + \frac{v_\emptyset}{r\sin\emptyset \sin\theta}\frac{\partial v_\emptyset}{\partial \emptyset} + \frac{v_\emptyset v_r - v_\theta v_\emptyset \cot\emptyset \cot\theta}{r}\right)$$

$$= -\frac{1}{r\sin\emptyset \sin\theta}\frac{\partial p}{\partial \emptyset} - \left[\frac{1}{r^3}\frac{\partial}{\partial r}(r^3\tau_{r\emptyset}) + \frac{1}{r\sin\emptyset \sin\theta}\frac{\partial}{\partial \theta}(\tau_{\theta\emptyset}\sin\emptyset \sin\theta)\right.$$

$$\left. + \frac{1}{r\sin\emptyset \sin\theta}\frac{\partial}{\partial \emptyset}\tau_{\emptyset\emptyset} + \frac{(\tau_{\emptyset r} - \tau_{r\emptyset}) + \tau_{\emptyset\theta} - \cot\emptyset \cot\theta}{r}\right] + \rho g_\emptyset$$

(2.14)

[c] These equations have been written without making the assumption that τ is symmetric. This means, for example, that when the usual assumption is made that the stress tensor is symmetric, $\tau_{r\theta} - \tau_{\theta r} = 0$.

2.2.3 The Equation of Energy in Terms of q

$$[\rho\hat{C}_p DT/Dt = -(\nabla \cdot \mathbf{q}) - (\partial \ln \rho/\partial \ln T)_p Dp/Dt - (\tau : \nabla \mathbf{v})] \quad (2.15)$$

Cartesian coordinate (x,y,z):

$$\rho\hat{C}_p\left(\frac{\partial T}{\partial t} + v_\lambda\frac{\partial T}{\partial x} + v_y\frac{\partial T}{\partial y} + v_z\frac{\partial T}{\partial z}\right) = -\left[\frac{\partial q_\lambda}{\partial x} + \frac{\partial q_y}{\partial y} + \frac{\partial q_z}{\partial z}\right] - \left(\frac{\partial \ln \rho}{\partial \ln T}\right)_p$$

$$\frac{Dp}{Dt} - (\tau : \nabla \mathbf{v}) \quad (2.16)$$

Cylindrical coordinate (r,θ,z):

$$\rho\hat{C}_p\left(\frac{\partial T}{\partial t} + v_r\frac{\partial T}{\partial r} + \frac{v_\theta}{r}\frac{\partial T}{\partial \theta} + v_z\frac{\partial T}{\partial z}\right) = -\left[\frac{1}{r}\frac{\partial}{\partial r}(rq_r) + \frac{1}{r}\frac{\partial q_\theta}{\partial \theta} + \frac{\partial q_z}{\partial z}\right]$$

$$-\left(\frac{\partial \ln \rho}{\partial \ln T}\right)_p\frac{Dp}{Dt} - (\tau : \nabla \mathbf{v}) \quad (2.17)$$

Spherical coordinate (r,θ,Ø):

$$\rho\hat{C}_p\left(\frac{\partial T}{\partial t} + v_r\frac{\partial T}{\partial r} + \frac{v_\theta}{r}\frac{\partial T}{\partial \theta} + \frac{v_\emptyset}{r\sin\theta}\frac{\partial T}{\partial \emptyset}\right) = -\left[\frac{1}{r^2}\frac{\partial}{\partial r}(r^2 q_r)\right.$$

$$\left. + \frac{1}{r\sin\theta}\frac{\partial}{\partial \theta}(q_\theta\sin\theta) + \frac{1}{r\sin\theta}\frac{\partial q_\emptyset}{\partial \emptyset}\right] - \left(\frac{\partial \ln \rho}{\partial \ln T}\right)_p\frac{Dp}{Dt} - (\tau : \nabla \mathbf{v})$$

(2.18)

^{a} The viscous dissipation term, $-(\tau : \nabla \mathbf{v})$, may usually be neglected, except for systems with very large velocity gradients. The term containing $(\partial \ln \rho / \partial \ln T)_p$ is zero for fluids with constant ρ

These forms of FPE are very general, and for supercritical systems, we need to simplify them. Most SFE systems operate in cylindrical geometry. In the next chapter, these models will be simplified and the most common model for SFE application will be presented.

2.3 Hybrid Modeling or Gray Box

There are generally three approaches for building mathematical models: White box modeling or first principle modeling, where everything is considered to be known from physical laws. In black box modeling, all knowledge derives from measurements. These models are based on input and output data from the system. Artificial neural networks and statistical modeling are examples of black box modelings. In gray box modeling, both physical laws and observed measurements are used to design a model. Figure 2.1 depicts the conceptual difference between these three types of testing. Gray box assumes that the structure of the model is given from physical laws as a parameterized function, and the model parameters are obtained using the observed data (measurements) information (Davoody et al. 2012)

White box performs better in figuring out the knowledge in comparison with the black box. Dealing with black box demands less expertise, and also is simpler rather than white box in achieving the desirable result. When the process is complex to be modeled with a first principle modeling approach, or when execution time for FPM or there are some unknown parameters for developing the FPM, in these cases black box modeling is the common modeling approach. Black box models are easy to develop and usually are fast

Figure 2.1 Conceptual difference among three types of testing.

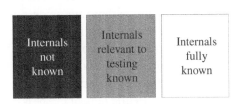

responding, but usually, there is doubt about their extrapolation capabilities (Zahedi et al. 2005).

Gray box analysis combines white box techniques with black box modeling. Gray box approaches usually require using several tools together. In fact, gray box mixes both black and white boxes in a fascinating way, black box modeling can scan systems across networks and white box modeling requires an equation to analyze the behavior statically. Grey box models use advantage of both first principle and black box models.

The main advantage of using gray box is combined benefits. Since this model is a combination of black box and white box testing, it presents advantages of both.

There are many hybrid-modeling formalisms, but the combination of first principle models with neural networks is a popular modeling paradigm (Kumar and Venkateswarlu 2012). The first principle models describes certain characteristics of the process being simulated and involves a multilayer feed forward neural networks, that serve as estimator of unmeasured process parameters that are difficult to model from first principles (Azarpour et al. 2015; Guo et al. 2001).

Zahedi et al. (2005) described that the hybrid first principle-neural network model may be developed in various combinations. The common approach is having the first principle model as the basis, while the neural network calculates the unknown parameter. Another approach is using the neural network to learn the deviation between mathematical model output and the target output. Another possible alternative is using the deterministic model as reinforcement for the function relationship between the inputs and the outputs.

The integration of the process knowledge from the first principle model into the neural networks makes the combined model to serve as an estimator of unmeasured process parameters that are difficult to be determined from the first principle model alone. The neural network provides a mapping between input and output through a nonlinear function. The first principle model provides process variables interactions from the system. The hybrid model is able to interpolate and extrapolate much more accurately, easier to analyze and interpret, and requires significantly fewer training samples compared to standard black box neural network models.

The common configurations of hybrid models are the first principle model hybrid with neural network in series and parallel. Serial approaches force the output from the neural network to be consistent with the first principle model, while parallel approaches will use the principle model to assist the network. Figure 2.2 shows the block diagram of these two approaches. Psichogios and

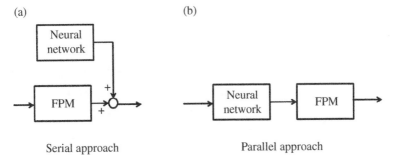

Figure 2.2 Two approaches used in hybrid neural network with first principle.

Ungar (1992) were among the first who proposed the serial approach for hybrid modeling of bioprocess and stated that the hybrid approach provides better prediction than black box model with pure neural network approach.

2.4 ANN

In 1943 Warren McCulloch, a neurophysiologist, and Walter Pitts, an Massachusetts Institute of Technology (MIT) logician introduced the artificial neurons. Artificial neurons play the role of processing elements in ANN, which can demonstrate difficult acts. ANNs learn by example, like people, and they can be considered as a nonlinear system that is useful for a variety of problems such as pattern recognition, data classification, and function approximation (Bulsari 1995).

Neurons, in fact interconnected processing elements, are the main members of ANN which bear the burden of solving problems. Between the neurons, there are synaptic connections that need to be adjusted. This is a part of the learning process of biological systems. We have the same condition for ANNs. ANN is able to target and find the similarities among various input types after it is trained (Yang et al. 2009).

2.4.1 Simple Neural Network Structure

A simple neural network structure is a perceptron, a classifier. This ANN was introduced in 1957 at the Aeronautical Laboratory by Frank Rosenblatt. It performs based on a feedforward method. Figure 2.3 shows a single-layer perceptron.

56 | 2 Modeling and Optimization Concept

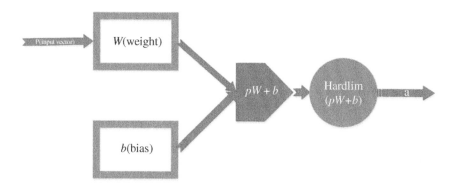

Figure 2.3 Single-layer perceptron.

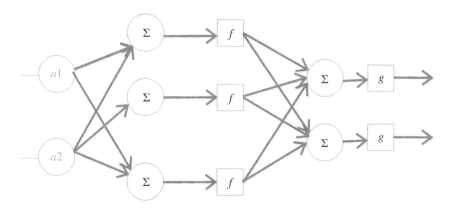

Figure 2.4 Two-layer perceptron.

A two-layer network is shown in Figure 2.4. Strength of a synapse in this figure is shown by the weight (w), the cell body is represented by the summation and the transfer function, f. The monitoring equation for the application of the network is $y = f(wp+b)$, in which f refers to transfer function, w stands for weight vector, p refers to input vector, and b corresponds to bias. The role of transfer function is played by neurons. The result of parameters inside the bracket is known as n which is the net input. Net result can be summarized in Eq. 2.19:

$$\text{net} = x_1 w_1 + x_2 w_2 + x_3 w_3 = \sum_{i=1}^{3} x_i w_i \qquad (2.19)$$

As the input vector **p** is multiplied by the weight vector **w**, it forms **wp**, terms which are sent to summers. The other input multiplied by bias will be moved to summer. Result is **n** which moves to transfer function, creates the final output vector (a^1) for layer one. This vector, a^1, will be used for the next layer as an input vector and a similar procedure as before, will produce output vector (a^2).

$$a_1 = f(w_1 p + b_1) = f(n_1) \qquad (2.20)$$

Important point to notice is that weight vector and bias are both adjustable scalar parameter adjustments of parameters w and b is done by learning rules of chosen transfer function.

2.4.1.1 Transfer Function
Transfer functions and weights play the most important roles in the learning algorithms of ANN. Transfer functions are divided into three types:

1) linear (or ramp)
2) threshold
3) sigmoid

In linear units, the output activity is proportional to the total weighted output. For threshold units, the output is set at one of two levels, depending on whether the total input is greater than some threshold value or less. For sigmoid units, the output varies continuously but not linearly as the input changes. In addition, sigmoid units bear a greater resemblance to real neurons than do linear or threshold units, but all three must be considered rough approximations. There are different transfer functions in ANN development (Hagan et al. 1995).

2.4.1.2 Activation Functions
There are four main types of activation functions Tan-sigmoid, Log-sigmoid, Hard Limit, and symmetrical Hard Limit (Figure 2.5). Different activation functions affect the performance of an ANN.

2.4.1.3 Learning Rules
Upgrading of **w** and **b** is done by a procedure known as learning rule (or training algorithm). Learning rules are categorized in three groups: supervised learning, unsupervised learning, and reinforcement learning. Supervised learning: examples of net behavior are given to learning rule:

$$\{p_1, a_1\}, \{p_2, a_2\}, ..., \{p_n, a_n\}$$

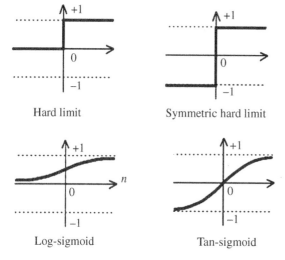

Figure 2.5 Common activation functions.

Here p_n refers to the input vector to the network, and a_n stands for real output vector or target. As inputs enter the network, one process is done to the inputs, resulting in outputs reached by the network. Variation between the reached output and real output (target) is observed. According to the variation among output (result of network) and target (the result which network is supposed to reach), an error is calculated. Then, the learning rule upgrades the weights and biases of the network in a way that reduces calculated error (Bulsari 1995).

No target is available in unsupervised learning. The only way to modify weights and biases is to deal with network inputs. Most algorithms follow clustering functions. They divide the input data into specific classes (Lang 2001).

If we provide the algorithm with only a grade instead of correct output or target corresponding to each network input, supervised learning shifts to reinforcement learning. The grade or scope is a tool by which the performance of a network is evaluated. It seems this learning is suitable for control system applications (Quah and Ng 2007).

2.4.2 Network Architecture

While dealing with different problems, most of the time, one neuron cannot be effective and maybe more numbers of neurons are needed operating in parallel, which is known as a "layer." Regarding more difficult problems, we need to use network architectures such as single-layer perceptron or multilayer perceptron (MLP) (Hagan et al. 1995). Single-layer perceptron is shown in Figure 2.6.

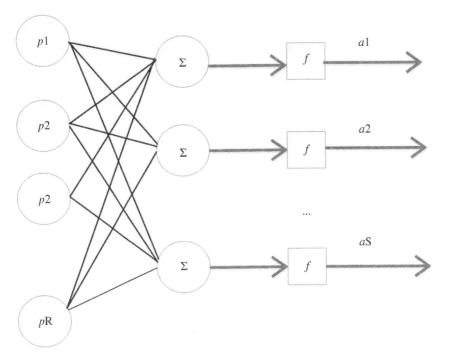

Figure 2.6 Layer of S neurons.

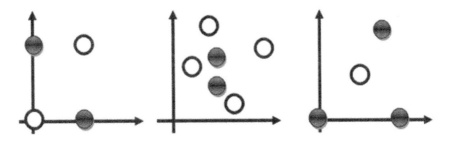

Figure 2.7 Linearly inseparable problems.

Single-layer networks have a limitation that this network is applicable to categorize input vectors which are able to be divided to two regions by linear boundary. Such vectors are known as "linearly separable," and unfortunately, many problems are not linearly separable (Bishop 1995; Heaton 2015). A few of such examples are shown in Figure 2.7.

Nowadays, MLP is the most popular network architecture. The network follows a form of input–output model, having the weights and biases. The units perform a biased weighted sum of their inputs, which results in output in feedforward topology. They are able to simulate functions of arbitrary complexity, by number of layers and units of each one, determining the function complexity. Important issues in multilayer perceptron design include specification of hidden layers' numbers and the number of units in these layers. A typical MLP is shown in Figure 2.8.

MLP uses back propagation techniques for learning and upgrading the values of weights and biases. Network needs a set of input/output examples. After the first process is done, output is reached based on specific input data, and difference between output and target determines the error. Once the error is calculated, the network starts to upgrade the weight and bias values in order to reduce the difference between the next coming output and its target. Due to this backwarding movement to upgrade the values, this technique is named as backpropagation. In this technique, correction of values start from the last layer where output is reached, and then it moves back through the previous layers to apply the new values to them. After repeating this for a specific time, the network gets acceptable error. There is also another technique for upgrading weight and bias which is known as gradient descent. For this, the derivative of the error based on the weights is determined, and the weights are shifted in order to reduce the error. It is highly used for nonlinear data (Yang et al. 2009). Table 2.1 lists some of algorithms and the acronyms used to identify them. All of them apply the gradient of the performance function (net.performFcn) to decide how to fix w to reduce performance (TheMathWorksInc. 2010).

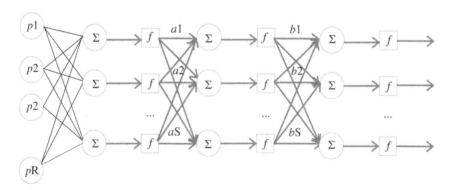

Figure 2.8 Multilayer perceptron (*Source:* Modified from Hagan et al. (1995). © John Wiley & Sons).

Table 2.1 List of algorithms.

Acronym	Algorithm	
LM	trainlm	Levenberg–Marquardt
BFG	trinbfg	BFGs Quasi-Neqton
RP	trainrp	Resilient Backpropoagation
SCG	trainscg	Scaled Conjugate Gradient
CGB	traincgb	Conjugate Gradient with Powell/Beale Restarts
CGF	traincgf	Fletcher–Powell Conjugate Gradient
CGP	traincgp	Polak–Ribiere Conjugate Gradient
OSS	trainoss	One Step Secant
GDX	traingdx	Variable Learning Rate Backpropagation

Source: The MathWorks Inc.

2.5 Fuzzy Logic

In 1965, Lotfi Zadeh introduced fuzzy logic through a famous article (Zadeh 1973). He developed his idea to a system of mathematical logic. Fuzzy logic refers to the new logic for showing fuzzy terms created by Lotfi Zadeh. The basis of this idea arises from human communication. Fuzzy logic tries to simulate our reactions, our behavior, our decision-making, and our common sense (Zadeh 1965). Fuzzy logic performs well as it deals with significance and precision. One obvious feature of fuzzy logic is its simple mathematical concepts. Basically, the idea of this theory believes that everything such as height, pressure, sweetness, darkness, and color can belong to different groups at the same time (by having a defined degree of membership) (Zadeh 1965). For example, the weather is really humid or Ali is a very tall guy.

2.5.1 Boolean Logic and Fuzzy Logic

Boolean logic follows a simple idea: True or False, Belongs or Doesn't. There is no middle ground for any data in this logic. Data are supposed to belong to only one group at time, and there is specific separation between data which belong and data which do not.

Fuzzy logic presents knowledge by a set of mathematical principles. These principles rely on degrees of membership. In contrast with Boolean logic which has two outcome values (0 or 1), fuzzy logic offers multiple values for data. What is more, it works on degrees of truth or membership (Běhounek and Cintula 2006).

Fuzzy logic employs a range of values between 0 to 1 in order to demonstrate the amount of membership of any data due to any fuzzy set. In contrast with Boolean logic which has only two colors, Fuzzy logic demonstrates the percentage of truth by spectrum of colors, suggesting that data would be partly true and partly false simultaneously.

2.5.2 Fuzzy Sets

The logic behind crisp set theory allocates only one value to any member in regard of membership: 1 or 0, white or black. The logic fails to represent vague concepts, and as a result of that, is incapable to respond to the paradoxes. In fuzzy set theory which is based on fuzzy logic, any element receives a degree of membership (Zadeh 1973). The degree is chosen from a real number in the interval [0, 1]. Following is an instance for comparing two logics and corresponding sets in the real world.

This example is about a tall fuzzy set (Table 2.2). The elements or data of this set are all women, but they have different degrees of membership.

Table 2.2 Degree of membership for height example.

Name	Height (cm)	Degree of membership	
		Crisp	Fuzzy
Raha	208	1	1
Lili	205	1	1
Helena	198	1	0.98
Atty	181	1	0.82
Helya	179	0	0.78
Sara	172	0	0.24
Neda	167	0	0.15
Rastin	158	0	0.06
Saba	155	0	0.01
Rose	152	0	0.00

2.5.3 Membership Function

The main task of membership function is to decide by which degree any element belongs to the fuzzy set. In other words, it allocates membership degrees to elements in order to determine how much the element belongs to the related fuzzy set as shown in Figure 2.9 (Zadeh 1973).

2.5.3.1 Membership Function Types

In this part, the fuzzy sets of the "height" example are demonstrated. Women can participate in three groups of short, average, and tall. It can be obviously seen in Figure 2.10 that according to classical set theory, the woman with height of 183 cm is a tall woman and belongs only to a tall group, while fuzzy

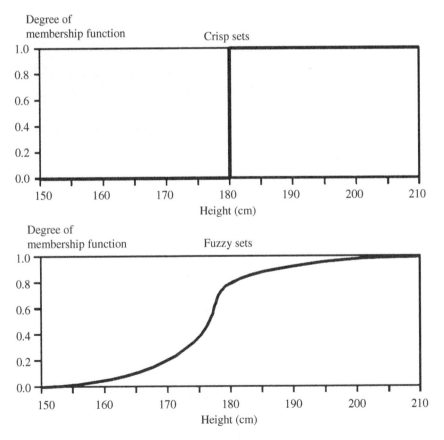

Figure 2.9 Schematic degrees of membership function for crisp and fuzzy sets (*Source:* Modified from Zadeh (1973). © John Wiley & Sons).

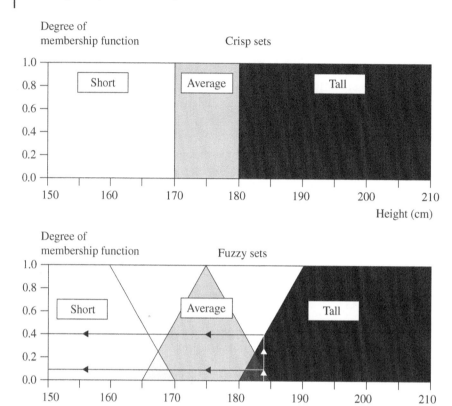

Figure 2.10 Determination of the membership function.

membership function says this woman belongs to a tall and average group simultaneously with different degrees (she is 0.1 average and 0.4 tall). Different membership function types are shown in Figure 2.11.

2.5.4 Fuzzy Rules

The fuzzy rule is typically written as below:

IF x is A THEN y is B

In this statement x and y are linguistic variables, A and B are linguistic values determined by fuzzy sets on the universe of discourses X and Y, respectively (Zadeh 1973).

2.5 Fuzzy Logic

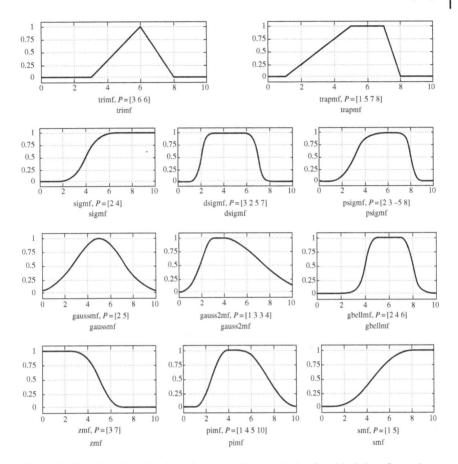

Figure 2.11 Membership function types (generated using Matlab software).

2.5.4.1 Classical Rules and Fuzzy Rules

Deep difference between these two types of rules is elaborated using an example:

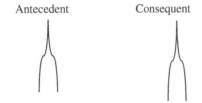

Antecedent   Consequent

Rule1: IF height is >180 THEN   the woman is tall

Rule2: IF height is < 180 THEN   the woman is not tall

Height can have any value in the range between 0 and 210 cm, but the linguistic variable "woman" is supposed to admit only value tall or not tall. As you can recall, classical rules follow the "black or white" idea of Boolean logic.

In fuzzy form, the height has specific range (0–200) that includes fuzzy sets. The woman should not be necessarily tall or short, but can be both. Elements have different fuzzy membership degrees to fuzzy sets, and fuzzy rules' role is to decide which set takes more members in comparison with others. After identifying the dominant fuzzy set, consequent becomes clear.

2.5.5 Fuzzy Expert System and Fuzzy Inference

Fuzzy inference is the process of formulating the mapping from a given input to an output using theory of fuzzy sets. Three main concepts including membership functions, fuzzy logic operators, and fuzzy if-then rules are utilized (Mamdani and Assilian 1975). Two famous types of fuzzy inference systems (FIS) are Mamdani type and Sugeno type.

2.5.5.1 Mamdani FIS

The Mamdani inference system includes five steps: Fuzzification, fuzzy logical operations, implication method, aggregation, and defuzzification (Mamdani and Assilian 1975).

2.5.5.1.1 Fuzzification

The fuzzification comprises the process of transforming crisp values into grades of membership for linguistic terms of fuzzy sets. At the first step, data enters the network, and the network clarifies the membership degree by which any inserted data belongs to the fuzzy sets. So, in this part, membership degrees of data are introduced to the system.

2.5.5.1.2 Fuzzy Logical Operation and Rule Evaluation

The input of this section is membership degrees. After receiving the degrees, fuzzy rules which are designed and defined by experts will be applied to the degrees in order to make a decision. For example, if the fuzzified data reveal that the temperature of the reactor is hot (antecedent), fuzzy rule commands that the temperature should reduce (consequent). This consequent, which is the output of this section, is shown by a simple value (most of the times numerical one) (Mamdani and Assilian 1975).

2.5.5.1.3 Implication Method
An implication method specifies how a fuzzy inference system scales the membership functions of an output linguistic variable based on the rule weight of the corresponding rule (Mamdani and Assilian 1975).

2.5.5.1.4 Aggregation of the Rule Outputs
In this part, all the outputs are required to become united. In order to do so, all membership functions are mixed and are represented as a fuzzy set. Therefore, inputs of this part are membership functions, and the output is a fuzzy set (Mamdani and Assilian 1975).

2.5.5.1.5 Defuzzification
The process of converting the fuzzy output is called defuzzification. At the beginning of the process, the values have fuzzified, and in the end, they have to be defuzzified in order to show the output as a crisp number. While it receives the output of the previous section (fuzzy set) as an input, the generated output is a single number. There are some different methods to defuzzify in Mamdani style such as COG (center of gravity). Centroid technique probably is the most popular one (Van Broekhoven and De Baets 2009):

$$COG = \frac{\int_a^b \mu_A(x) x \, dx}{\int_a^b \mu_A(x) \, dx} \tag{2.21}$$

2.5.5.2 Sugeno Fuzzy Inference
Mamdani style inference requires one to find the centroid of a two-dimensional shape by integrating across a continuously varying function. In general, this process is not computationally efficient. To solve this problem, Michio Sugeno (Takagi and Sugeno 1985) suggested using a single spike, a singleton, as the membership function of the rule consequent. A singleton, or more precisely a fuzzy singleton, is a fuzzy set with a membership function that is unity at a single particular point on the universe of discourse and zero everywhere else. The format of the Sugeno style fuzzy rule is

IF x is A AND y is B THEN z is f (x, y)

where x, y, and z are linguistic variables, A and B are fuzzy sets on the universe of discourses X and Y, respectively. And $f(x, y)$ is a mathematical function (Figure 2.12).

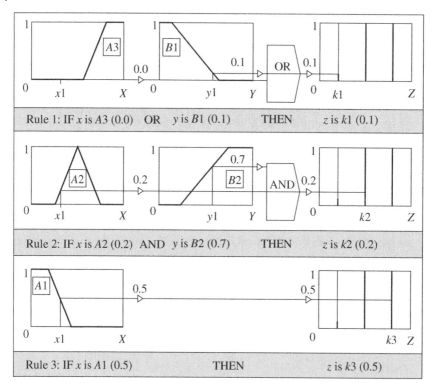

Figure 2.12 Sugeno-style rule evaluation.

2.6 Neuro Fuzzy

Fuzzy logic and neural networks are natural complementary tools in building intelligent systems. While neural networks are low-level computational structures that perform well when dealing with raw data, fuzzy logic deals with reasoning on a higher level, using linguistic information acquired from domain experts. However, fuzzy systems lack the ability to learn and cannot adjust themselves to a new environment. On the other hand, although neural networks can learn, they are opaque to the user (Hocaoglu et al. 2009).

Neuro fuzzy system is a hybrid intelligent system which combines two intelligent technologies, neural network and fuzzy logic. Integrated neuro fuzzy systems can combine the parallel computation and learning abilities of neural networks with the human-like knowledge representation and explanation abilities of fuzzy systems. As a result, neural networks become more transparent, while fuzzy systems become capable of learning (Zahedi et al. 2013).

2.6.1 Structure of a Neuro Fuzzy System

The structure of a neuro fuzzy system is similar to a multilayer neural network. In general, a neuro fuzzy system has input and output layers, and three hidden layers that represent membership functions and fuzzy rules (Figure 2.13).

Each layer in the neuro fuzzy system is associated with a particular step in the fuzzy inference process. First layer is the input layer where each neuron transmits external crisp signals directly to the next layer. Second layer is the fuzzification layer that neurons in this layer represent fuzzy sets used in the antecedents of fuzzy rules. In addition, a fuzzification neuron receives a crisp input and determines the degree to which this input belongs to the neuron's fuzzy set. Next layer is the fuzzy rule layer. Each neuron in this layer corresponds to a single fuzzy rule. A fuzzy rule neuron receives inputs from the fuzzification neurons that represent fuzzy sets in the rule antecedents. The fourth layer is the output membership function layer. Neurons in this layer represent fuzzy sets used in the consequence of fuzzy rules, and the last layer is the defuzzification layer. Each neuron in this layer represents a single output of the neuro fuzzy system. It takes the output fuzzy sets clipped by the respective integrated firing strengths and combines them into a single fuzzy set (Hocaoglu et al. 2009).

2.6.2 Adaptive Neuro Fuzzy Inference System (ANFIS)

ANFIS works in Takagi–Sugeno-type fuzzy inference systems (Sugeno and Kang 1988), which was developed by Jang (1993), Jang and Sun (1995) and

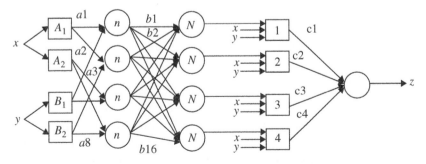

Figure 2.13 Structure of a neuro fuzzy system.

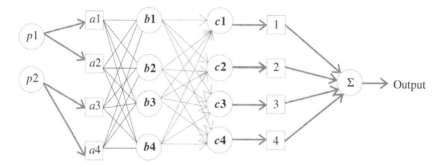

Figure 2.14 Structure of ANFIS.

Jang et al. (1997). The most common inference system used in ANFIS is a first-order Sugeno-type FIS which is in the form of relation. During the training procedure, rule parameters including antecedent parameters and consequent parameters will be tuned to present more accurate outputs with the minimum error. There is no need to know any background knowledge of rule parameters. ANFIS can learn the parameters and demonstrate membership functions. Structure of ANFIS is similar to that of multilayer feed forward ANN (Jang 1993). The architecture of ANFIS is shown in Figure 2.14, and the neuron function in each layer is described below.

Layer 1: Neurons in the first layer are known as input layers and do not perform specific operations on the inputs but just pass them off to the next layer.

Layer 2: This layer is the fuzzification layer. Fuzzification was performed by neurons in this layer. Fuzzification neurons have a bell membership function, in Jang's model.

Layer 3: Or rule layer in which every neuron refers to a single Sugeno-type fuzzy rule. Neurons in this layer get the output of the previous layer and evaluate the firing strength of the rule it represents (Potter and Negnevitsky 2006). In ANFIS, the product operator is used to evaluate the conjunction of the rule antecedents. Thus, the output of neuron i in this layer is calculated by

$$y_i^{(3)} = \prod_{j=1}^{k} x_{ji}^{(3)} \qquad y_{\Pi^1}^{(3)} = \mu_{A1} \times \mu_{B1} = \mu_1 \qquad (2.22)$$

Layer 4: This layer is the normalization layer. In this layer, neurons get the output of the previous layer and evaluate the normalized firing strength of the rule. The normalized firing strength is the ratio of the firing strength of a given rule to the sum of firing strengths of all rules. Equations below demonstrate the output of this layer:

$$y_i^{(4)} = \frac{x_{ii}^{(4)}}{\sum_{j=1}^{n} x_{ji}^{(4)}} = \frac{\mu_i}{\sum_{j=1}^{n} \mu_j} = \overline{\mu}_i \quad y_{N1}^{(4)} = \frac{\mu_1}{\mu_1 + \mu_2 + \mu_3 + \mu} = \underline{\mu}_1$$

(2.23)

Layer 5: This layer is the defuzzification layer. Each neuron in this layer is connected to the respective normalization neuron and also receives initial inputs, x_1 and x_2. A defuzzification neuron computes the weighted consequent value of a given rule as follows:

$$y_i^{(5)} = x_i^{(5)}[c_{i0} + a_{i1}x_1 + b_{i2}x_2] = \underline{\xi}_i[c_{i0} + a_{i1}x_1 + b_{i2}x_2]$$ (2.24)

where x_i^5 is the input and y_i^5 is the output of defuzzification neuron i in Layer 5, and c_{i0}, a_{i1} and b_{i2} is a set of consequent parameters of rule i.

Layer 6: This layer is represented by a single summation neuron. This neuron computes the sum of outputs of all defuzzification neurons and generates the overall ANFIS output, y,

$$y = \sum_{i=1}^{n} x_i^{(6)} = \sum_{i=1}^{n} \underline{\xi}_i[c_{i0} + a_{i1}x_1 + b_{i2}x_2]$$ (2.25)

2.6.2.1 Learning in the ANFIS Model

ANFIS can use either a back propagation form of the steepest descent method for membership function parameter estimation, or a hybrid which uses a combination of back propagation and the least-squares method to estimate membership function parameters. In the Sugeno style fuzzy inference, an output, y, is a linear function. Thus, given the values of the membership parameters and a training set of P input–output patterns, P linear equations in terms of the consequent parameters will be formed as follows:

$$\{y_d(1) = \underline{\mu}_1(1)f_1(1) + \underline{\mu}_2(1)f_2(1) + \ldots + \underline{\mu}_n(1)f_n(1)\,y_d(2)$$

$$= \underline{\mu}_1(2)f_1(2) + \underline{\mu}_2(2)f_2(2) + \ldots + \underline{\mu}_n(2)f_n(2) \vdots y_d(p)$$

$$= \underline{\mu}_1(p)f_1(p) + \underline{\mu}_2(p)f_2(p) + \ldots + \underline{\mu}_n(p)f_n(p) \vdots y_d(P)$$

$$= \underline{\mu}_1(P)f_1(P) + \underline{\mu}_2(P)f_2(P) + \ldots + \underline{\mu}_n(P)f_n(P)$$

In the matrix notation, have

$$y_d = A\,k,$$ (2.26)

where, y_d is a $P \times 1$ desired output vector

$$y_d = \begin{bmatrix} y_d(1) \, y_d(2) \vdots y_d(p) \vdots y_d(P) \end{bmatrix} \tag{2.27}$$

$$A = [\underline{\mu}_1(1)\underline{\mu}_1(1)x_1(1) \ldots \underline{\mu}_1(1)x_m(1) \ldots \underline{\mu}_n(1)\underline{\mu}_n(1)x_1(1) \ldots \underline{\mu}_n(1)x_m(1)$$

$$\underline{\mu}_1(2)\underline{\mu}_1(1)x_1(2) \ldots \underline{\mu}_1(2)x_m(2) \ldots \underline{\mu}_n(2)\underline{\mu}_n(2)x_1(2) \ldots \underline{\mu}_n(2)x_m(2) \vdots$$

$$\underline{\mu}_1(p)\underline{\mu}_1(p)x_1(p) \ldots \underline{\mu}_1(p)x_m(p) \ldots \underline{\mu}_n(p)\underline{\mu}_n(p)x_1(p) \ldots \underline{\mu}_n(p)x_m(p) \vdots$$

$$\underline{\mu}_1(P)\underline{\mu}_1(P)x_1(P) \ldots \underline{\mu}_1(P)x_m(P) \ldots \underline{\mu}_n(P)\underline{\mu}_n(P)x_1(P) \ldots \underline{\mu}_n(P)x_m(P) \,]$$
$$\tag{2.28}$$

And k is an n (1 + m) ×1 vector of unknown consequent parameters,

$$k = [k_{10}k_{11}k_{12}\ldots k_{1m}k_{20}k_{21}k_{22}\ldots k_{2m}\ldots k_{n0}k_{n1}k_{n2}\ldots k_{nm}] \tag{2.29}$$

When the rule consequent parameters are identified, we can calculate the actual network output vector, y, and also the error vector, e:

$$e = y_d - y \tag{2.30}$$

In the backward pass, the back-propagation algorithm is applied. The error signals are propagated back, and the antecedent parameters are updated according to the chain rule, while the consequent parameters are kept fixed. The hybrid-learning algorithm has been used in all ANFIS structures because it is highly efficient in training (Tahmasebi and Hezarkhani 2012).

2.7 Optimization

Human being is faced with decision-making in his daily life and during this decision-making process he tries to make the best decisions. Mathematics has played an important role in decision-making and providing optimal answers for human beings. Using experience also is an alternative solution to find optimal answers but sometimes using experience needs devoting time, energy, and also costly. For example, the best time for waking up has been obtained by experience but finding minimum production cost to produce a product needs spending money and time. In such cases, employing optimization techniques can save us time and money. In order to form optimization usually an objective function is needed to be optimized (minimized or maximized). In engineering, sometimes this objective function can be maximizing profit, minimizing costs, maximizing production, minimizing wastes, and many other objective functions can be considered.

After forming objective functions, some constraints may exist in the process. The goal is finding optimal points, but in this case, some variables need to be varied. For example, inlet temperature feed flow rate can change for SFE to find optimum condition. But in practice, there is a limit for temperature. Maximum available temperature is known based on process limitation. Also, the minimum temperature can be a critical temperature of a solvent to keep it in supercritical condition. The flow rate range depends on pump or compressor speed range.

Having these types of constraints help us to limit our search domain and reach the optimum point soon. After defining the constraint, selection of optimization methods is important. There is not any generic framework for selection of optimization techniques. Generally, the optimization techniques are divided into two main categories: gradient based (traditional) techniques and evolutionary (nontraditional) techniques. In the following these two techniques will be described, but the general rule is that if the objective function and constraints can be defined as explicit functions, traditional optimization methods can be employed. If the objective function and constraints cannot be described by explicit functions, or they be very complex functions, in these conditions nontraditional optimization methods are used.

2.7.1 Traditional Optimization Methods

These methods are based on gradient search methods, for example in a simple case consider finding roots of an equation. In this case, by finding a point which makes the function positive and one point with negative answer, it is estimated that the root is between these two points. Based on function sign, the search domain is narrowed until the root with reasonable approximation is found, but maybe there are other roots also. Optimization gradient-based rule needs that the user makes sure the obtained optimum is not local minimum and it is global minimum (Fig. 2.15). Also, the functions need to be continuous in these methods and speed and convergence of gradient methods highly depend on initial guess.

In linear programming methods, there is no need for continuity of the functions and the optimum value which exists for a matrix of equation nothing is seek. The equations can be nonlinear and, in this case, instead of continuity, convexity of the equations is important. It guarantees having global minimum.

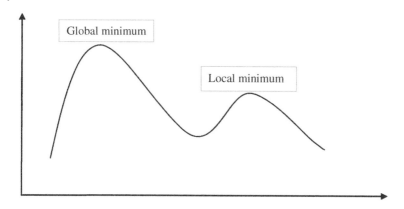

Figure 2.15 Global and local optimums.

2.7.2 Evolutionary Algorithm

Using gradient methods in optimization problems has some difficulties. For complex models, the computational time is long. Sometimes calculating differentials is impossible, and sometimes the functions are not continuous. The answer depends on initial guesses, and there is a chance of having local optimums instead of global optimums (Fogel 2000).

Evolutionary algorithms are very helpful in the cases where the classic gradient-based optimization techniques fail to provide an answer. There are many evolutionary algorithms such as genetic algorithms, ant colonies, and simulated annealing and more algorithms also are developing. Mother Nature is the inspiration for most of these algorithms. The way schools of fish swim or how an offspring is generated or how ants can carry food are inspiration for some of these algorithms. We only describe three algorithms in this book, but there are more evolutionary algorithms to be explored.

2.7.3 Simulated Annealing Algorithm

This algorithm was developed by Metropolice in 1953 (Eglese 1990). The algorithm is based on cooling of a metal. During the cooling process, solid atoms are located near each other in a way that when the metal reaches thermal equilibrium, the atoms have minimum energy. This algorithm starts with an initial random configuration. Then new configurations are developed and are compared with each other. Change in the objective function which is the energy of the system is criteria for abandoning old configuration and adapting new configuration. At the beginning, there are many configurations, but as the temperature drops, the chance of choosing wrong configuration decreases. The Boltzmann activation energy function is used to check the possibilities (Henderson et al. 2003).

2.7.4 Genetic Algorithm

Usually, solving optimization problems includes the development of models for the objective function (using mathematical approaches such as regression models, theoretical analysis models, or differential equations) and the determination of optimal conditions which are searched using a search method (such as direct search, grid search, or gold section method for a single variable, or alternating variable search, pattern search or Powell's method for multiple variables). With rapid development in computer programs, artificial intelligence technologies such as ANNs and GAs have been used for optimization of nonlinear and complex systems (Chen and Ramaswamy 2002).

GA is an optimization technique which is inspired by the biological evolutionary processes (selection, crossover, and mutation). This procedure searches the optimum value in parallel.GA minimizes the objective function by changing the variable values randomly not according to its derivatives. Because of its stochastic identification, sticking in the local minimum is a small possibility which is a good advantage of this technique, but sometimes, the process speed may become slow due to limitation on some of its parameters properly (Bollas et al. 2004).

Heavy oil demand has decreased due to energy conservation and nuclear power development, and the demand for light products has increased. This raises the problem of the conversion of heavy products into light products such as gasoline, kerosene, and diesel fuel for the production of motor fuels (Sheikhattar et al. 2011).

2.7.4.1 Genetic Algorithm Definitions

Genetic Algorithms (GA) were invented to mimic some of the processes observed in natural evolution. Many people, biologists included, are astonished that life at the level of complexity that we observe could have evolved in the relatively short time suggested by the fossil record. The idea with GA is to use this power of evolution to solve optimization problems. The father of the original GA was John Holland who invented it in the early 1970s.

A GA is an iterative procedure maintaining a population of structures that are candidate solutions to specific domain challenges. During each temporal increment (called a generation), the structures in the current population are rated for their effectiveness as domain solutions, and on the basis of these evaluations, a new population of candidate solutions is formed using specific genetic operators such as reproduction, crossover, and mutation (Renner and Ekárt 2003). They combine survival of the fittest among string structures with a structured yet randomized information exchange to form a search algorithm with some of the innovative flair of human search. In every generation, a new set of artificial creatures (strings) is created using bits and pieces of the fittest of the old; an occasional new part is tried for good measure. While randomized, genetic algorithms are no simple random walk. They efficiently exploit historical information to speculate on new search points with expected improved performance.

2.7.4.2 Genetic Algorithms Overview

GAs are search algorithms based on the mechanics of the natural selection process (biological evolution). The most basic concept is that the strong tend to adapt and survive while the weak tend to die out. That is, optimization is based on evolution, and the "Survival of the fittest" concept. GAs have the ability to create an initial population of feasible solutions, and then recombine them in a way to guide their search to only the most promising areas of the state space. Each feasible solution is encoded as a chromosome (string) also called a genotype, and each chromosome is given a measure of fitness via a fitness (evaluation or objective) function. The fitness of a chromosome determines its ability to survive and produce offspring. A finite population of chromosomes is maintained.

GAs use probabilistic rules to evolve a population from one generation to the next. The generations of the new solutions are developed by genetic recombination operators:

- Biased reproduction: selecting the fittest to reproduce.
- Crossover: combining parent chromosomes to produce children chromosomes.
- Mutation: altering some genes in a chromosome.

Crossover combines the "fittest" chromosomes and passes superior genes to the next generation. Mutation ensures the entire state-space will be searched, (given enough time) and can lead the population out of a local minima. Most important parameters in GAs are the following:

- Population size
- Evaluation function

- Crossover method
- Mutation rate

Determining the size of the population is a crucial factor. Choosing a population size too small increases the risk of converging prematurely to a local minima, since the population does not have enough genetic material to sufficiently cover the problem space. A larger population has a greater chance of finding the global optimum at the expense of more CPU time. The population size remains constant from generation to generation (Holland 1975).

2.7.4.3 Preliminary Considerations

1) Determine how a feasible solution should be represented
   a) Choice of alphabet. This should be the smallest alphabet that permits a natural expression of the problem.
   b) The string length. A string is a chromosome and each symbol in the string is a gene.
2) Determine the population size. This will remain constant throughout the algorithm. Choosing a population size too small increases the risk of converging prematurely to a local optimum, since the population does not sufficiently cover the problem space. A larger population has a greater chance of finding the global optimum at the expense of more CPU time.
3) Determine the objective function to be used in the algorithm.
4) Determine an initial population (i) random or (ii) by some heuristic
5) REPEAT

    A) Determine the fitness of each member of the population

    Perform the objective function on each population member. Fitness scaling can be applied at this point. Fitness scaling adjusts down the fitness values of the super-performers and adjusts up the lower performers, promoting competition among the strings. As the population matures, the really bad strings will drop out. Linear scaling is an example.

    B) Reproduction (Selection)

    Determine which strings are "copied" or "selected" for the mating pool, and how many times a string will be "selected" for the mating pool. Higher performers will be copied more often than lower performers. Example: the probability of selecting a string with a fitness value of f is f/ft, where ft is the sum of all of the fitness values in the population.

C) Crossover

Mate each string randomly using some crossover technique 2. For each mating, randomly select the crossover positions (Note one mating of two strings produces two strings. Thus, the population size is preserved).

D) Mutation

Mutation is performed randomly on a gene of a chromosome. Mutation is rare, but extremely important. As an example, perform a mutation on a gene with probability 0.005. If the population has g total genes (g = string length population size), the probability of a mutation on any one gene is 0.005 g, for example. This step is a no-op most of the time. Mutation ensures that every region of the problem space can be reached. When a gene is mutated, it is randomly selected and randomly replaced with another symbol from the alphabet UNTIL Maximum number of generations is reached.

2.7.4.4 Overview of Genetic Programming

Genetic programming starts with an initial population of randomly generated computer programs composed of functions and terminals appropriate to the problem domain. The functions may be standard arithmetic operations, standard programming operations, standard mathematical functions, logical functions, or domain-specific functions. Depending on the particular problem, the computer program may be Boolean-valued, integer-valued, real-valued, complex-valued, vector-valued, symbolic valued, or multiple-valued. The creation of his initial random population is, in effect, a blind random search of the search space of the problem. Each individual computer program in the population is measured in terms of how well it performs in the particular problem environment. This measure is called the fitness measure. The nature of the fitness measure varies with the problem (Koza and Koza 1992).

Koza's initial problem: Given a set of initial predicates and possible actions, develop (evolve) a computer program (in Lisp) to control the movement of an ant searching for food. The chromosome is a variable sized Lisp program where the leaf nodes are actions (left, right, move, etc.), and the internal nodes are predicates or logic controls (if found food), etc. Each chromosome (program) is used to control the actions of a simulated ant in searching for food. The evaluation function for a given chromosome is the amount of food gathered by an aunt in a fixed amount of time.

2.7.4.5 Implementation Details

Based on natural selection, after an initial population is randomly generated, the algorithm evolves through three operators: Selection which equates to the survival of the fittest; crossover which represents mating between individuals; and mutation which introduces random modifications.

2.7.4.5.1 Selection Operator

- Key idea: give preference to better individuals, allowing them to pass on their genes to the next generation.
- The goodness of each individual depends on its fitness.
- Fitness may be determined by an objective function or by a subjective judgment.

2.7.4.5.2 Crossover Operator

- Prime distinguished factor of GA from other optimization techniques.
- Two individuals are chosen from the population using the selection operator see Fig. 2.16.
- A crossover site along the bit strings is randomly chosen.
- The values of the two strings are exchanged up to this point.
- If S1 = 000000 and s2 = 111111, and the crossover point is 2, then S1'=110000 and s2'=001111.
- The two new offspring created from this mating are put into the next generation of the population.
- By recombining portions of good individuals, this process is likely to create even better individuals.

2.7.4.5.3 Mutation Operator

- With some low probability, a portion of the new individuals will have some of their bits flipped.
- Its purpose is to maintain diversity within the population and inhibit premature convergence depicted in Fig. 2.17.
- Mutation alone induces a random walk through the search space.

Mutation and selection (without crossover) create a parallel, noise-tolerant, hill-climbing algorithm (Michell 1998).

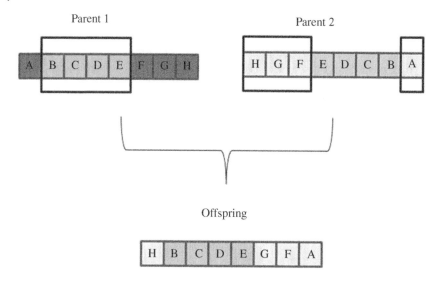

Figure 2.16 Schematic of crossover operation in GA.

2.7.4.6 Effects of Genetic Operators

- Using selection alone will tend to fill the population with copies of the best individual from the population.
- Using selection and crossover operators will tend to cause the algorithms to converge on a good but suboptimal solution.
- Using mutation alone induces a random walk through the search space.
- Using selection and mutation creates a parallel, noise-tolerant, and hill climbing algorithm.

2.7.4.7 The Algorithms

1) Randomly initialize population (t)
2) Determine fitness of population (t)
3) Repeat
    a) select parents from population (t)
    b) perform crossover on parents creating population ($t + 1$)
    c) perform mutation of population ($t + 1$)
    d) determine fitness of population ($t + 1$)
4) Until best individual is good enough

Figure 2.17 Schematic of the mutation operation in GA.

Before mutation

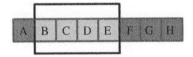

After mutation

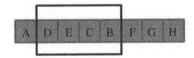

In the previous subsection, it has been claimed that via the operations of selection, crossover, and mutation, the GA will converge over successive generations toward the global (or near global) optimum. Why this simple operation should produce fast, useful, and robust techniques is largely due to the fact that GAs combine direction and chance in the search in an effective and efficient manner. Since populations implicitly contain much more information than simply the individual fitness scores, GAs combine the good information hidden in a solution with good information from another solution to produce new solutions with good information inherited from both parents, inevitably (hopefully) leading toward optimality.

The ability of the algorithm to explore and exploit simultaneously, a growing amount of theoretical justification, and successful application to real-world problems strengthens the conclusion that GAs are a powerful, robust optimization technique.

3
Physical Properties of Palm Oil as Solute

3.1 Introduction

When considering refining processes and edible oil fractionation, it is first necessary to examine the characteristics and composition of the oil. With this information, it is possible to select the most suitable and practical refining methods as well as process conditions and the manner in which the oil can be used as end products. In this chapter, palm oil's physical properties as well as chemical properties that were derived from experimental studies are presented. This chapter ends with a detailed description of the available current technologies for the removal of undesirable components and the recovery of desirable components from palm oil through physical refining and chemical methods.

3.2 Palm Oil Fruit

Palm oil is produced from the fruit of oil palm trees (Elaeis Guineensis) which originated in West Guinea. Most of the world's production of palm oil comes from South-East Asia, in particular, Malaysia and Indonesia. These countries are the largest contributors, with the respective average up to 40–50% production of palm oil globally. As one of the two world's biggest producers and exporters of palm oil and palm oil products, Malaysia plays an important role in fulfilling the growing global needs for oils and fats.

The average weight of each bunch varies between 10 and 30 kg, and individual fruits are generally in the range of 8–20 g. The individual fruit is made up of an outer skin (the exocarp), a pulp (mesocarp) containing the palm oil in a fibrous matrix, a central nut consisting of a shell (endocarp), and the kernel

Modeling, Simulation, and Optimization of Supercritical and Subcritical Fluid Extraction Processes,
First Edition. Zainuddin A. Manan, Gholamreza Zahedi, and Ana Najwa Mustapa.
© 2022 by the American Institute of Chemical Engineers, Inc.
Published 2022 by John Wiley & Sons, Inc.

(the seed), which itself contains an oil, quite different from palm oil, resembling coconut oil (Sundram et al. 2003).

A unique feature of the oil palm is that it produces two types of oil – palm oil that is extracted from mesocarp, and palm kernel oil that is extracted from the palm kernel. Palm oil is rich in carotenoids (pigments found in plants and animals) from which it derives its deep red colour. The major component of its triglycerides is palmitic acid, which is a saturated fatty acid which exists as a viscous semisolid under ambient tropical condition and becomes solid fat in temperate climates.

Palm oil is an important source of food and a major source of lipid. Steady increase in the world's population increases the demand for palm oil as an important source of edible oils and fats. Almost 90% of the world palm oil production is traded as edible oils and fats. At 75.45 million metric tons of palm oil produced in the 2020/2021 crop year, palm oil accounts for 36% of the total world production of oil and fats and has overtaken soybean oil as the world's most important vegetable oil. (Statista, 2021). This phenomenon arises due to the unique characteristics of palm oil, particularly its potential health benefits. The viability of palm oil for export is determined by the ability of the oil palm to be grown successfully. High yields of the oil palm throughout the year are essential to meeting the high global market demand.

3.3 Palm Oil Physical and Chemical Properties

Vegetable oil extracted from fleshy mesocarp of palm fruit is called crude palm oil (CPO). The oil is a semisolid material at a room temperature and has a melting point of 36 °C. In principle, palm oil composition determines the oil's chemical and physical characteristics. The palm oil composition comprises of triglycerides (95%) as the major components, diglycerides (2–7%) and monoglycerides (1%), free fatty acids (3–5%), and minor components (1%) (such as tocopherol, carotenoids, phosphatides, aliphatic alcohols, and sterols (Choo et al. 2005; Kumar and Krishna 2014). Palm oil glycerides are made up of a range of fatty acids with 50% saturated fatty acid and 50% unsaturated fatty acid (Efendy Goon et al. 2019). The major saturated fatty acids comprise of palmitic acid (45%) and stearic acid (>3.5%), while the unsaturated fatty acids are mainly oleic acid (>40%) and linoleic acid (>10%) (Japir et al. 2017). Figure 3.1 shows the typical constituents of crude palm oil.

A good quality CPO is measured by the content of the free fatty acids (FFA). High content of FFA (>5%) in CPO is considered as poor-quality CPO (MPOB

3.3 Palm Oil Physical and Chemical Properties

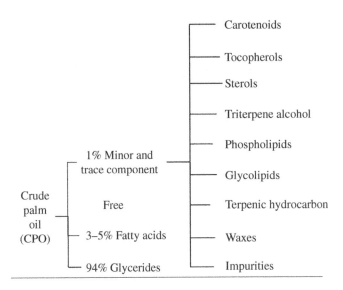

Figure 3.1 Constituents of crude palm oil (*Source:* Goh et al. (1987). © John Wiley & Sons).

2004) for food application and only suitable for nonedible industries such as biodiesel, candle, and bioplastic industries. The high content of FFA is caused by the degradation of lipase in the crude palm oil due to hydrolyzation of triglycerides in the seeds. This usually happens if the fruits are stored for a lengthy period. On the other hand, low content of FFA (<5%) indicates an excellent CPO quality; therefore, suitable for food and oleochemical applications. In Malaysia, it is approximated that about 80% of the palm oil produced are of excellent quality.

3.3.1 Palm Oil Triglycerides

Most of the fatty acids in palm oil are present as triglycerides (TG) consisting primarily of palmitic acid and oleic acid. The composition of the palm oil lipid class such as fatty acids, triacylglycerols, and polar lipids change over the development period of the palm mesocarp, as shown in Table 3.1 (Oo et al. 1986).

The palm oil triglycerides partially define most of the physical characteristics of the palm oil such as melting point and crystallization behavior. The different placement of fatty acids and fatty acid types on the glycerol molecule produces a number of different triglycerides. There are 7–10% of saturated triglycerides, predominantly tripalmitin and 6–12% of fully unsaturated

Table 3.1 Changes in fatty acid composition of lipid classes at different stages of development of oil palm mesocarp.

Weeks after anthesis	Lipid class	Percentage of total fatty acids[a]									
		10:0	12:0	14:0	16:0	18:0	18:1	18:2	18:3	20:3	22:0
8	Total lipids	4.20	3.10	1.00	27.50	4.40	22.20	24.00	13.60		
12		1.40	1.50	1.00	27.00	4.50	22.70	23.90	18.00		
16		0.10	1.70	0.40	35.20	5.40	42.60	13.90	0.80		
20		0.40	5.50	1.10	40.80	5.00	35.90	11.30	0.00		
Overripe		0.00	0.80	1.50	44.20	5.40	38.70	9.40	0.00		
8	Triacylglycerols	0.00	0.70	1.40	43.20	3.20	18.90	20.60	10.80	1.20	
12		0.10	0.70	1.30	25.80	1.40	24.50	25.70	20.60	0.00	
16		0.40	0.80	0.60	38.90	4.80	42.60	11.90	0.00	0.00	
20		0.00	0.10	1.40	45.40	5.20	36.70	11.10	0.00	0.00	
Overripe		0.00	0.30	1.40	56.30	3.20	33.30	5.60	0.00	0.00	
8	Fatty acids	0.00	1.40	2.30	49.80	2.60	20.30	7.00	8.70	7.80	
12		0.00	3.50	3.00	25.40	0.40	45.00	9.50	1.40	11.90	
16		0.00	0.60	0.00	28.30	0.80	54.90	15.40	0.00	0.00	
20		0.20	0.20	2.50	58.80	6.00	24.70	7.60	0.00	0.00	
Overripe		0.00	0.20	1.50	35.40	0.50	45.20	17.20	0.00	0.00	
8	1,3-Diacylglyoerols	0.00	33.30	7.30	52.40	3.50	3.40	0.00			
12		1.00	18.00	8.20	42.40	5.20	25.50	0.00			

16		0.10	2.10	0.60	68.10	5.80	21.90	1.80		
20		1.10	0.90	2.80	51.30	5.90	29.70	8.30		
Overripe		0.00	0.20	1.90	55.80	4.60	31.20	6.20		
8	1,2-Diacylglycerols	0.20	1.40	2.20	74.20	4.10	7.50	0.00	0.00	10.40
12		0.10	1.30	1.70	62.50	3.60	12.80	13.60	3.00	1.30
16		0.00	0.20	0.10	35.70	1.90	45.40	16.70	0.00	0.00
20		0.30	0.00	1.30	41.80	4.60	37.50	14.50	0.00	0.00
Overripe		0.00	0.40	1.60	55.50	3.40	31.40	7.80	0.00	0.00
8	Monoacylglycerols	0.20	0.40	3.90	60.00	3.80	14.00	7.60	2.30	3.40	4.30
12		0.10	1.70	1.80	44.30	2.90	14.70	22.90	11.70	0.00	0.00
16		0.00	0.60	0.80	66.30	8.50	18.50	5.30	0.00	0.00	0.00
20		0.70	1.80	1.40	67.00	9.70	14.80	4.70	0.00	0.00	0.00
Overripe		0.10	2.20	2.00	59.10	5.60	28.80	7.20	0.00	0.00	0.00
8	Polar lipids	0.00	0.30	0.50	40.90	2.00	14.00	28.60	13.70	0.00
12		0.00	0.30	0.60	32.50	2.10	17.20	24.80	22.30	0.00
16		0.20	1.00	0.30	33.40	1.90	24.50	28.00	10.80	0.00
20		1.30	0.30	1.20	45.60	4.30	18.90	14.80	13.60	0.00
Overripe bonds.		0.50	2.00	1.40	40.80	0.70	33.40	14.60	5.40	1.10

[a] Fatty acids are denoted by the number of carbon atoms: the number of double bonds.
Source: Oo et al. (1986). © Elsevier.

triglycerides. A typical Malaysian refined palm oil has a triglycerides carbon number profile as shown in Table 3.2. The typical fatty acid composition of palm oil is presented in Table 3.3. Palm oil has saturated and unsaturated fatty acids in approximately equal amounts.

Table 3.2 Triglycerides composition (wt %) of Malaysian palm oil.

Carbon no.	Mean	Range observed	Standard deviation	Coefficient of variation (%)
C44	0.07	0–0.2	0.06	84.8
C46	1.18	0.7–2.0	0.19	16.7
C48	8.08	4.7–9.7	0.72	8.9
C50 (POP, PPO)	39.88	38.9–41.6	0.54	1.3
C52 (POO)	38.77	33.1–41.0	0.62	1.6
C54	11.35	10.3–12.1	0.37	3.3
C56	0.59	0.5–0.8	0.10	17.1

Source: Tan and Oh (1981). © John Wiley & Sons.

Table 3.3 Typical fatty acid composition of Malaysian palm oil.

Fatty acid chain length	Composition (wt %)	Range observed	Standard deviation
C12:0 (Lauric acid)	0.3	0–1.0	0.12
C14:0 (Myristic acid)	1.1	0.9–1.5	0.08
C16:0 (Palmitic acid)	43.5	39.2–45.8	0.95
C16:1 (Palmitoleic acid)	0.2	0–0.4	0.05
C18:0 (Stearic acid)	4.3	3.7–5.1	0.18
C18:1 (Oleic acid)	39.8	37.4–44.1	0.94
C18:2 (Linoleic acid)	10.2	8.7–12.5	0.56
C18:3 (Linolenic acid)	0.3	0–0.6	0.07
C20:0 (Arachidic acid)	0.2	0–0.4	0.16

Source: Tan and Oh (1981). © John Wiley & Sons.

3.3.2 Minor Components in Palm Oil

The minor constituents of palm oil can be divided into two groups. The first group consists of fatty acid derivatives, such as partial glycerides (mono- and diacylglycerols), phosphatides, esters, and sterols. The second group includes classes of compounds not related chemically to fatty acids. These are the hydrocarbons, aliphatic alcohols, free sterols, tocopherols, pigments, and trace metals. Table 3.4 shows the levels of these minor components in the oil. The groups of most importance to refiners are the carotenoids, tocopherols, and phosphatides.

The partial glycerides do not occur naturally in significant amounts except in palm oil from damaged fruits. Such oils would have undergone partial hydrolysis resulting in the production of FFA, water, and the partial glycerides. Phospholipids are better known to the refiner as phosphatides and are frequently

Table 3.4 Contents of various components in the unsaponifiable fraction for Malaysian palm oil.

	Component	Percentage (%)	mg/kg (in palm oil)
Carotenoids	α-carotene	29	500–700
	β-carotene	62	
	γ-carotene	4	
	Lycopene	2	
	Xanthophylls	3	
Tocopherols	α-tocopherol	20	~800
	α-tocotrienol	25	
	γ-tocotrienol	45	
	δ-tocotrienol	10	
Sterols	Cholesterol	4.1	326–627
	Campesterol	22.8	
	Stigmasterol	11.3	
	β-sitosterol	57.5	
Phosphatides		98	20–80

Source: Modified from Sambanthamurthi et al. (2000). © John Wiley & Sons.

referred to, together with small quantities of glycosides, carbohydrates, and pectins, as "gums," which have adverse effects on product quality and yield of refined oil.

One of the uniqueness of palm oil is its high content of carotenoids. Typical crude palm oil contains 500–700 ppm of carotenoids. The major carotenoids in palm oil are β- and α-carotene which account for 90% of the total carotenoids. Carotenoids are the precursors of vitamin A, with β-carotene having the highest pro-vitamin A activity. They are thermally decomposed during the refining process. Figure 3.2 shows the degree and rate of decomposition of β-carotene at various temperatures.

Vitamin E is a fat-soluble vitamin, which comprises two major homologous series of compounds (tocochromanols), known as tocopherols and tocotrienols. Vegetable oils, especially palm oil, are rich sources of tocopherols. The vitamin E content in CPO ranges between ~800 parts per million (ppm) and is a mixture of tocopherols (18–22%) and tocotrienols (78–82%).

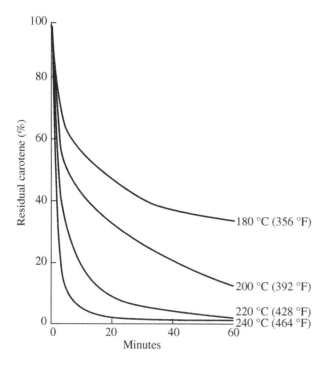

Figure 3.2 Thermal destruction of β-carotene (*Source:* Gunstoner (1987). © John Wiley & Sons).

Together with tocopherols, carotenoids contribute to the stability and nutritional value of palm oil. Therefore, the extraction and recovery of the carotenes would give a significant added value to the oil. Unfortunately, these valuable carotenes are destroyed and removed during conventional palm oil refining to give a light-colored oils as required by consumers, rendering them unavailable for recovery and use (Gast et al. 2001; Ooi et al. 1996).

Besides, conventional processing of palm oil also leaves high content of carotenoids in pressed palm fibers that have been found to be a good source of carotene. The residual palm fibers from palm oil production contain between 4000 and 6000 ppm of carotenoids, six times higher than that found in crude palm oil (Franca and Meireles 1997). The residue is a waste product which is typically burned in palm oil mills, regardless of the high carotenoid contents (Birtigh et al. 1995; Franca and Meireles 1997).

A number of methods have been developed to recover and extract carotenoids from crude palm oil including solvent extraction, adsorption, precipitation, and transesterification-distillation. However, transesterification-distillation and transesterification-solvent extraction have been scaled-up to industrial practise. These methods are energy-intensive processes since the esters must be vacuum-distilled. In such conditions, carotenes may undergo thermal degradation.

3.4 Vegetable Oil Refining

Crude oils obtained by oilseed processing have to be refined before the consumption in order to remove undesirable accompanying substances. High-quality products are characterized by low FFA content. Several physical procedures have been developed to produce edible oils with very low acid values. An overview of methods of crude vegetable oil refining is given in Table 3.5. To date, steam refining is the only large-scale practicable method used in the industry to remove FFA from crude oils.

3.5 Conventional Palm Oil Refining Process

Palm oil and palm kernel oil consist of mainly glycerides and small variable portions of nonglycerides components in crude form as well. In order to render the oils in their edible form, some of these nonglycerides need to be either removed or reduced to acceptable levels. The nonglycerides are of two broad

Table 3.5 Methods of refining of crude vegetable oils.

Procedure	Principle
Steam refining (Physical refining)	Removal of free fatty acids and other volatiles by superheated steam at 200–270 °C at low pressure after preliminary degumming and bleaching steps (Bloemen 1966; Koseoglu et al. 1998)
Alkali refining (Chemical refining)	Free fatty acids are neutralized by washing crude oils with a solution of sodium hydroxide or sodium carbonate followed by bleaching and deodorization process (Klein 1981)
Inert gas stripping	Removal of free fatty acids in a stream of inert gas (nitrogen) (Cheng et al. 1993; Fernandez Vecilla 1994; Krishnamurthy et al. 1992)
Molecular distillation	Removal of more volatile components, including FFA, from less volatile triacylglycerols at very low pressure, without application of steam (Choo et al. 1997; Rüütmann and Kallas 1994)
Membrane refining	Treatment of crude oils under pressure with selective membranes, permeable of free fatty acids, but not permeable for triacylglycerols (Koseoglu et al. 1996)
Extraction with solvents	Selective removals of fatty acids by countercurrent distribution between immiscible solvents, one dissolving selectively free acids while another one selectively extracting triacylglycerols (Raman et al. 1996; Sahashi et al. 1994)
Refining with supercritical CO_2	Removal of free fatty acids and other impurities using supercritical carbon dioxide (Ooi et al. 1996; Ziegler and Liaw 1993)

types: oil insoluble and oil soluble. The insoluble impurities consisting of fruit fibers, nut shells, and free moisture mainly are readily removed. The oil-soluble nonglycerides are more difficult to remove and thus, the oil needs to be subjected to various stages of refining. They include free fatty acids, phospholipids, trace metals, carotenoids, tocopherols and tocotrienols, oxidation products, and sterols.

Not all of the nonglycerides components are undesirable. The tocopherols and tocotrienols not only help to protect the oil from oxidation, which is detrimental to flavor and keep-ability of finished oil, but also have nutritional attributes. α- and β-carotene present as major constituents of carotenoids are precursors of vitamin A. The other impurities are generally detrimental to the oil's flavor, odor, color, and keep ability and thus influence the oil's usefulness.

The basic function of the refining process is to remove FFA, moisture and impurities, all oxidative products and carotene from the CPO. The resultant oil termed as refined, bleached, and deodorized palm oil (RBDPO) contains FFA content of less than 0.1%. Currently, CPO is refined through physical or chemical refining. Chemical refining includes degumming, neutralization, bleaching, and deodorization. Physical refining, which is the more popular and cheaper technique, is based on the higher volatility of FFA compared to triglycerides. In physical refining, removal of FFA by chemical neutralization is replaced by simultaneous deacidification and deodorization. Typical refining process consists of three major process sections, namely: degumming, bleaching and deodorization, as explained in the following section.

3.5.1 Chemical Refining

Before the start of the offtake from the crude oil tank, the oil undergoes heating up to the required temperature (about 45 °C) to ease pumping and keeps homogenized to help affect a final product consistency. The crude oil then undergoes gum conditioning (Figure 3.3). The crude oil is pumped through a heat exchanger to be heated up to temperature of about 80 °C. The oil is then treated with 0.05–0.10% food-grade orthophosphoric acid in a mixer. A reaction time of 15 minutes is allowed where gum (phosphatides) is made easily removable at the next stage.

The acid treated oil is then continuously dosed with caustic soda. The concentration and amount of alkali used will vary with the FFA content of the oil. The alkali reacts with the FFA forming precipitated soaps, which are removed either through centrifuge or settling and washing. The light phase discharge is mainly refined oil containing traces of soap and moisture, while the heavy phase discharge primarily contains soap, insoluble materials, gums, free alkali, and minute quantities of neutralized oil. Certain amount of neutral oil is saponified along with the FFA and is lost by emulsification.

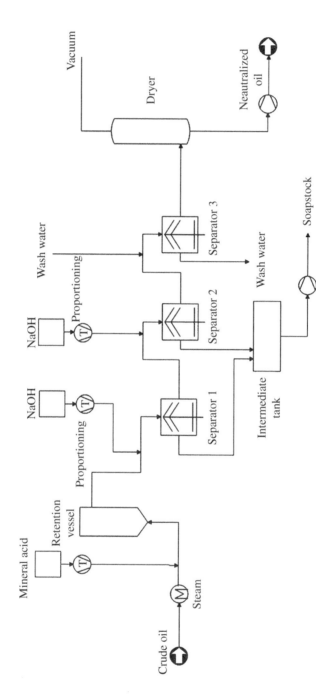

Figure 3.3 Chemical refining – degumming and neutralization steps.

3.5 Conventional Palm Oil Refining Process

After the above steps of phosphoric acid treatment for gum removal and neutralization for FFA reduction, the oil still contains undesirable impurities (odors and color pigments) that need to be removed before the finished product will be acceptable. Some of these remaining impurities are removed in quantity by the process of bleaching or adsorptive cleansing (Figure 3.4). The practice of bleaching involves addition of activated clay (bleaching earth) and always improves the initial taste, final flavor, and oxidative stability of the product. It helps to overcome problems in subsequent processing by adsorption of soap traces, prooxidant metal ions, decomposes peroxides, and adsorbs other minor impurities. Bleaching is carried out under vacuum at a temperature of about 100 °C and given a reaction time of half an hour. The dosage of earth varies with type and quantity of starting oil and is usually in the range of 0.5–1.0%. As mentioned earlier, the primary function of bleaching earth is to reduce undesirable impurities through adsorption. However, a certain amount of bleaching (color reduction) by pigment adsorption occurs as a bonus effect. Color reduction is actually effected in the next stage through high temperature thermal destruction of pigments.

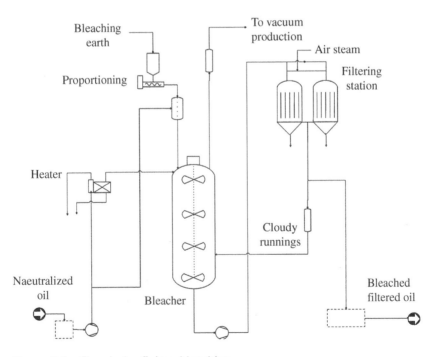

Figure 3.4 Chemical refining–bleaching.

The slurry containing the oil and earth is then passed through the main filter to give clear, free-from-earth particles oil. Usually, a second check filter is used in series with the main filter to double ensure no earth slips occur. Presence of earth fouls deodorizers, reduces the oxidative stability of product oil, and acts as catalyst for dimerization and polymerization activities. Some amount of oil is lost through entrapment in waste earth, and it is usually in the order of 20–45% of the weight of dry earth.

The neutralized, bleached oil then proceeds to the next stage where the FFA content and color are further reduced and it is deodorized to produce a product, which is stable and bland in flavor (Figure 3.5). Deodorization is basically a high temperature, high vacuum, steam distillation process. A deodorizer operates in the following manner: de-aerates the oil, heats up the oil, steam strips the oil, and cools the oil before it leaves the system.

Deodorization can be carried out in batch, continuous, or semicontinuous style. The present practices in Malaysia are to go for the more efficient and less-cost continuous and semicontinuous processes. In a continuous alkali-refining route, the oil is generally heated to 220–240 °C under vacuum. A vacuum of 2–5 mbar is usually maintained through the use of ejectors

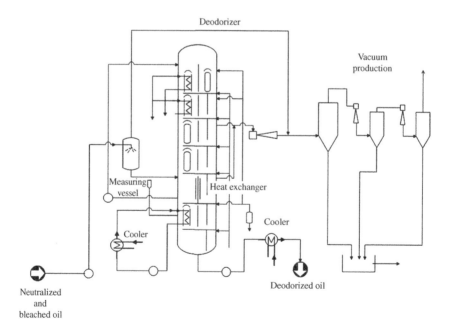

Figure 3.5 Chemical refining – deodorization.

and boosters. Heat bleaching of the oil occurs at this temperature through thermal destruction of carotenoid pigments. The use of direct stripping steam ensures the removal of residual FFA, aldehydes, and ketones which are responsible for unacceptable odors and flavors.

The oil leaves the deodorizer still under vacuum and cooled down to less than 60 °C. It passes through a polishing filter before it is sent to the storage tank. The oil is now termed neutralized, bleached, and deodorized (NBD) palm oil.

3.5.2 Physical Refining

Physical refining of CPO is the more common process in Malaysia for reasons of higher efficiency, less losses, less operating costs, less capital input, and negligible effluent handling. The pretreatment stage of physical refining is exactly the same as the alkali route. Once again, phosphoric acid is used. At the bleaching earth addition stage, however, relatively higher doses of earth are used. The "excess" earth is used to adsorb impurities which are removed with soap stock and by washing via the chemical route. Earth dosage used is usually 1–2%. The filtered bleached oil is termed Degummed Bleached (DB) oil (Figure 3.6).

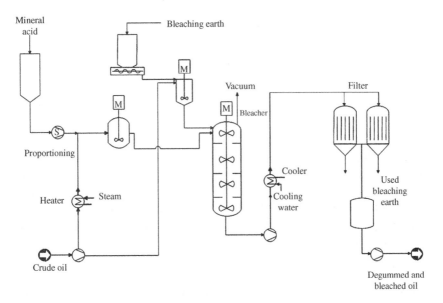

Figure 3.6 Physical refining – degumming and bleaching.

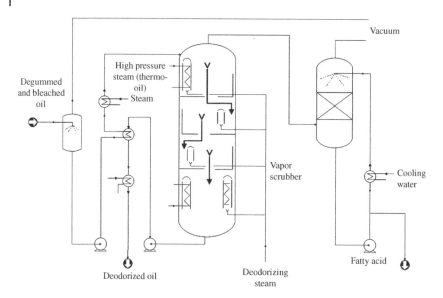

Figure 3.7 Physical refining – deacidification (by distillation) and deodorization.

The pretreated oil here enters the deodorizer at FFA content much higher than NB oil. As such, deodorization is of "heavy duty" using higher temperatures of 250–270 °C for palm oil and 240–245 °C for palm kernel oil, more stripping steam and thus a bigger vacuum system. The FFA distilled off is condensed and collected and are normally termed as palm fatty acid distillate (PFAD) (Figure 3.7).

3.5.3 Effect of Palm Oil Refining

The present refining practice for oil bleaching is through a series of destructive processes using either Fuller's earth, heating at about 240 °C or a combination of Fuller's earth and heating at about 150 °C. These processes destroy the carotenes in oil to produce a light yellow refined palm oil, and hence, represent the loss of a considerable potential source of natural vitamin A.

Thermal breakdown of tocopherols only occurs at temperatures above 260 °C. Under normally applied deodorization temperatures (220–260 °C), maximum 5% of the tocopherols are thermally degraded (De Greyt et al. 1999). The vitamin E in palm oil is mainly lost in the deacidification stage (typically 15–57%) of the palm oil refining process (Maclellan 1983). The vitamin E lost during processing is concentrated in the palm fatty acid distillate, a

Table 3.6 Specifications for crude palm oil and refined palm oil.

Specifications	Crude palm oil	Refined palm oil
Moisture (%)	0.15–0.20	0.1 max
Diglycerides (%)	±5	—
Monoglycerides (%)	0.2–0.3	—
Free fatty acids (%)	±4.5	0.1 max (as palmitic)
Phosphorus (mg/kg)	15–35	4 max
Carotene (mg/kg)	400–550	20

Source: Modified from Hui (1996). © John Wiley & Sons.

by-product of the physical refining of palm oil. This has been looked as a good source of tocopherol or vitamin E.

The refining processes remove FFA and deodorize the oil. However, it also reduces the tocopherol content and destroys all carotenes present in palm oil. Table 3.6 gives the specifications for the typical crude palm oil and refined palm oil.

A few methods were invented in order to preserve carotenoids and tocopherols. One example is a refining process at low temperature and high vacuum without addition of alkali to remove FFA (Chakrabarti and Jala 2019). The process operating conditions help the removal of FFA and peroxides products such as aldehydes, ketones, and trace pesticides residues with less destruction of the minor compounds. This technique is called a short-path distillation process that enables as high as 70% of carotenoids recovery (Sulihatimarsyila et al. 2019). A chemical deacidification using ethanol via continuous liquid–liquid extraction to remove FFA has been reported by Gonçalves et al. (2016). The FFA was successfully removed to as low as 0.3 wt% with high content of carotenoids recovered. However, the use of organic solvent is not preferable for food industries as it requires an energy and capital-intensive separation process to purify the deacidified palm oil.

A more advanced separation technique using supercritical carbon dioxide has been envisioned as considerably simpler and capable of producing high-quality refined palm oil while preventing the loss of carotenoids and tocopherols. Manan et al. (2009) simulated a new conceptual palm oil refining process using supercritical carbon dioxide. A two-stage process flowsheet to remove FFA and recover carotenoids and tocopherols was simulated using Aspen Plus software. A good agreement of the findings with pilot plant data showed the feasibility of the process to produce high-quality refined palm oil, and at the same time, retain the content of tocopherols and carotenoids.

3.6 Conclusions

Physical refining which consists of efficient degumming and deodorization (deacidification) steps has made great progress over the last 10 years. The physical refining procedures are commonly used for larger-scales production facilities. However, the choice of the optimum technology depends on the quality of the crude oils, processing equipment, tradition and experience, environmental and economical factors, and quality requirements for the refined products. Among these methods, supercritical fluid extraction technology has undergone rapid development and made promising progress in industrial applications.

4

First Principle Supercritical and Subcritical Fluid Extraction Modeling

Part I: Modeling Methodology

4.1 Introduction

In order to develop a supercritical or subcritical fluid extraction (SSFE) process, a reliable phase equilibrium prediction model for supercritical solvent–solute components systems, pure component property parameter, and models which can well represent the pure component properties as well as the extraction process have to be developed. In this chapter, a detailed methodology for the development of pure component property database for palm oil as well as the establishment of a reliable thermodynamic model for the palm oil component-supercritical CO_2 system is provided. Sections 4.2–4.7 describe the detailed modeling procedure for palm oil pure component property estimation, phase equilibrium calculations, and steady-state process simulation using a commercial process simulator, Aspen Plus® release 10.2.1. Results of pure component and mixture property modeling are presented in Sections 4.8–4.13. Results of base case process simulation are presented in Section 4.14.

4.2 Phase Equilibrium Modeling

The first step in modeling the phase equilibrium of a multicomponent palm oil-supercritical CO_2 system is to establish the composition of the multicomponent palm oil system (system characterization). This is followed by the prediction of palm oil components' solubilities and distribution coefficients using models such as the Redlich–Kwong–Aspen (RKA) equation of state. The RKA model can be applied to the SFE process since it is particularly suitable for modeling a mixture of polar components with light gases at medium-to-high pressures (Aspen Technology 2000). Prior to using the RKA-EOS to predict the

Modeling, Simulation, and Optimization of Supercritical and Subcritical Fluid Extraction Processes,
First Edition. Zainuddin A. Manan, Gholamreza Zahedi, and Ana Najwa Mustapa.
© 2022 by the American Institute of Chemical Engineers, Inc.
Published 2022 by John Wiley & Sons, Inc.

phase equilibria, some key physical and critical properties of the palm oil components were estimated and the binary interaction parameters of the equation of state were calculated.

4.3 The Redlich–Kwong–Aspen Equation of State

The RKA-EOS (Aspen Tech 2000) is a cubic equation of state that is an extension of the Redlich–Kwong–Soave equation of state (Soave 1972). The RKA-EOS was regressed using the Data Regression System module available in Aspen Plus® process simulator, version 10.2.1, to correlate the experimental phase equilibrium data published in the literature. Equation 4.1 represents the RKA-EOS used in this work for modeling the phase equilibria of the CPO-supercritical CO_2 system:

$$P = \frac{RT}{v-b} - \frac{a}{v(v+b)} \tag{4.1}$$

where P is pressure (in MPa), R is universal gas constant (8.314 J/mol/K), T is temperature (in K), v is molar volume (in m³/mol), a (in m⁶/MPa/mol) and b (in m³/mol) are the cross-energy and covolume parameters for a mixture.

4.3.1 Calculations of Pure Component Parameters for the RKA-EOS

For a pure component i, the parameters a_i and b_i for the RKA-EOS are functions of the critical temperature (T_{ci}) and critical pressure (P_{ci}) of the pure component:

$$a_i = \alpha_i 0.427\,47 \frac{R^2 T_{ci}^2}{P_{ci}} \tag{4.2}$$

$$b_i = 0.086\,64 \frac{RT_{ci}}{P_{ci}} \tag{4.3}$$

The RKA-EOS is not accurate in predicting vapor pressures below 10 torr (Sandler 1994). To improve the vapor pressure prediction for a highly nonlinear dependence of vapor pressure on temperature, Mathias (1983) recommends the generalized temperature-dependent function, α_i (see Eq. 4.4) for the subcritical component (referring to the palm oil components in this study) which considerably improves vapor pressure predictions:

$$\alpha_i(T) = \left[1 + m_i\left(1 - \sqrt{T_{ri}}\right) - \eta_i\left(1 - \sqrt{T_{ri}}\right)(0.7 - T_{ri})\right]^2 \tag{4.4}$$

where T_{ri} is the reduced temperature. The polar factor of the pure component i (η_i), which takes into account the polarity, is fitted from the pure component vapor pressure data. The constant for pure component i (m_i) is calculated as a function of the acentric factor (ω_i):

$$m_i = 0.480 + 1.574\omega_i - 0.176\omega_i^2 \tag{4.5}$$

4.3.2 Binary Mixture Calculations

The RKA-EOS utilizes the classical quadratic mixing rule for a mixture given in Eqs. 4.6 and 4.7. To model the molecular interactions between components i and j, the binary interaction parameters ($k_{a,ij}$, $k_{b,ij}$) were introduced through the quadratic mixing rules as follows:

$$a = \sum_i \sum_j x_i x_j \sqrt{a_i a_j}(1 - k_{a,ij}) \tag{4.6}$$

$$b = \sum_i \sum_j x_i x_j \frac{(b_i b_j)}{2}(1 - k_{b,ij}) \tag{4.7}$$

To predict the phase equilibrium of dissimilar components such as palm oil and supercritical CO_2, binary interaction parameters were required. To improve the predictive capability of the equation of state, the interaction parameters were considered temperature-dependent (Klein & Schulz 1989; Meier et al. 1994; Pereira et al. 1993). The RKA-EOS assumes a linear temperature dependency of the interaction parameters (Aspen Tech 2000) as follows:

$$k_{a,ij} = k^0_{a,ij} + k^1_{a,ij} \frac{T}{1000} \tag{4.8}$$

$$k_{b,ij} = k^0_{b,ij} + k^1_{b,ij} \frac{T}{1000} \tag{4.9}$$

4.4 Palm Oil System Characterization

Palm oil is essentially a mixture of saturated and unsaturated mixed triglycerides. It is not possible to know the exact distribution of the different fatty acid chains in the mixed triglyceride molecules. A description of complex natural systems may need a simple, although reasonable, hypothesis about the components involved. One possible approach is to represent the oil as a mixture of simple triglycerides (tripalmitin, triolein, etc.) in accordance with the fatty acid composition of the natural oil. Another alternative is to represent the oil by a single pseudo-component, which has the same molecular weight and degree of unsaturation of the original oil (Espinosa et al. 2002).

Researchers often treat the multicomponent-supercritical CO_2 mixture as a pseudo-binary system in which the main components in the oil are treated as a single component using a lumping procedure. In the study carried out by Mathias et al. (1986) on palm oil, the distribution of carbon atoms among the fatty acids and the occurrence of double bonds were ignored. Thus, the total number of saturated carbon atoms was used to designate the triglyceride components of the palm oil. Geana and Steiner (1995) used the same simplifying hypothesis to model the rapeseed oil. França and Meireles (2000) proposed a model that assumes the solutes extracted from pressed palm oil fibers as a mixture consisting of three key components, namely, oleic acid (as FFA), triolein (as TG), and β-carotene. Oghaki et al. (1989) used palmitic acid to represent the FFA, tripalmitin to represent TG, and α-tocopherol to represent vitamin E in the palm oil mixture.

Crude palm oil contains various components such as MG (<1%), DG (2–7%), TG (>90%), FFA (3–5%), phospholipids, and pigmented compounds, as well as several nutritionally bioactive compounds (Choo et al. 1996). Regarding the composition of CPO, the model proposed in this study considered CPO to be a mixture containing principally TG (tripalmitin and triolein) with some FFA (oleic acid) and minor components (β-carotene and α-tocopherol).

4.4.1 Palm Oil Triglycerides

Triglycerides are the major component in palm oil with the carbon number ranging mainly from C48 to C54, as given in Table 3.2. The fatty acid content in palm oil ranging from C12 to C22 with different degrees of saturation with palmitic acid and oleic acid as the major fatty acids in palm oil TG which constitute about 90% of the total TG (refer Table 3.3). The triglycerides in palm oil exist, mainly in the form of di-saturated dipalmitoyl-oleoyl-glycerol (POP and PPO) and monosaturated palmitoyl-dioleoyl-glycerol (POO). The physical data of these components are very scarce in the literature. Thus, simple TG, such as tripalmitin and triolein, which are relatively well being studied, were chosen to represent the saturated and unsaturated portions in palm oil TG, respectively. The structures of simple TG are presented in Figure 4.1.

The average molecular weight and the variance of palm oil TG were calculated based on the assumption that the distribution of fatty acids among the TG and the presence of unsaturation in fatty acids were negligible, as shown in Table 4.1. Owing to the relatively wide distribution of TG in palm oil, it was insufficient to represent the complex mixture of palm oil TG as a single

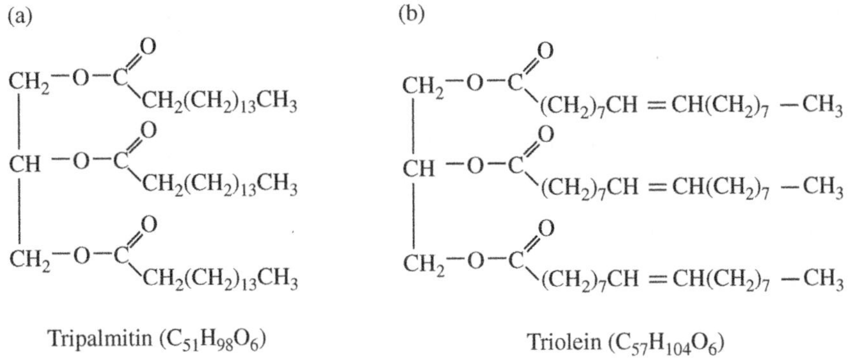

(a)

CH$_2$—O—C(=O)—CH$_2$(CH$_2$)$_{13}$CH$_3$
|
CH —O—C(=O)—CH$_2$(CH$_2$)$_{13}$CH$_3$
|
CH$_2$—O—C(=O)—CH$_2$(CH$_2$)$_{13}$CH$_3$

Tripalmitin (C$_{51}$H$_{98}$O$_6$)

(b)

CH$_2$—O—C(=O)—(CH$_2$)$_7$CH=CH(CH$_2$)$_7$—CH$_3$
|
CH —O—C(=O)—(CH$_2$)$_7$CH=CH(CH$_2$)$_7$—CH$_3$
|
CH$_2$—O—C(=O)—(CH$_2$)$_7$CH=CH(CH$_2$)$_7$—CH$_3$

Triolein (C$_{57}$H$_{104}$O$_6$)

Figure 4.1 The structure of simple triglycerides.

Table 4.1 Composition of triglycerides in palm oil.

Carbon number	Molecular weight (kg/kmol)	wt%
C44	750.9	0.07
C46	778.9	1.18
C48	806.9	8.08
C50	834.9	39.88
C52	862.9	38.77
C54	890.9	11.35
C56	918.9	0.59

MW = 848.9 kg/kmol; σ^2 = 592.78.
Source: Tan and Oh (1981). © John Wiley & Sons.

component. Saturated and unsaturated palm oil TG were represented by tripalmitin and triolein, respectively.

The composition of the selected TG components was based on the typical fatty acids composition in CPO (Tan and Oh 1981). The calculation of the composition for palm oil TG is included in Appendix A. In this study, the palm oil TG was approximated as a mixture of 48.8 wt% tripalmitin and 51.2 wt% triolein, with an average molecular weight of 846.7 g/mol.

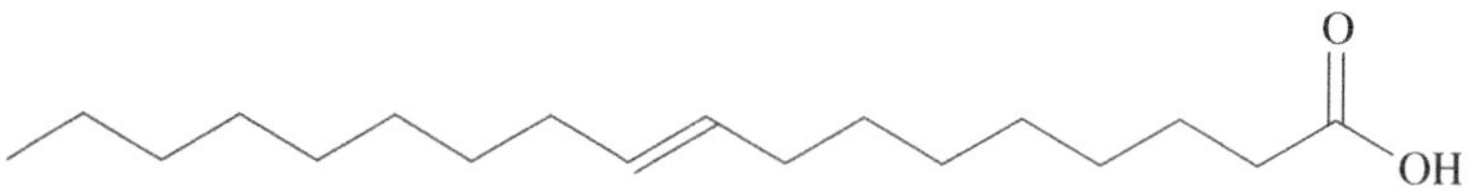

Figure 4.2 Chemical structure of oleic acid ($C_{17}H_{33}COOH$).

4.4.2 Free Fatty Acids

The free fatty acids (FFA) content of edible oil is one of the most useful quantitative indicators of oil quality. FFA results from the hydrolysis of TG, or more simply, oil is broken down to FFA by reacting with water. The breakdown of TG to FFA reduces oil quality. Palm oil, after extraction, contains approximately 4.5 wt% of FFA. Composition of the FFA present in CPO is not available in the open literature. A study on the changes in lipid class and composition in oil palm mesocarp by Sambanthamurthi et al. (2000) suggested that FFA consist of oleic acid (45.20 wt%), palmitic acid (35.40 wt%), and linoleic acid (17.20 wt%). According to Table 3.1, the composition of C18 fatty acids (oleic acid + linoleic acid) are essentially greater than C16:0. Therefore, oleic acid, as the most abundant component in FFA, was selected as the key component to represent the FFA in palm oil (Figure 4.2).

4.4.3 Palm Oil Minor Components

Palm oil contains mostly two isomers of tocopherols, α- and γ-tocopherol. Tocopherols and tocotrienols are present in crude palm oils at 600–1000 ppm levels. The α-tocopherol is the major activity isomer of vitamin E among four isomers α-, β-, δ-, and γ-tocopherols. In this study, α-tocopherol was selected to represent the tocopherols in palm oil. The structure of the α-tocopherol is presented in Figure 4.3:

The carotenoids content of palm oil varies between 500 and 700 ppm. A typical analysis of the carotenoids present in palm oil (given in Table 2.4)

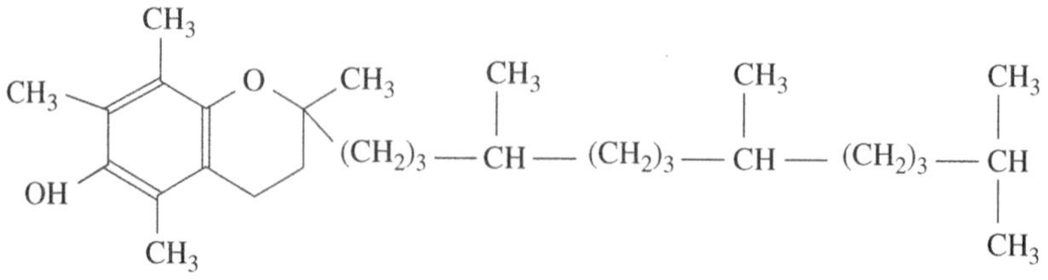

Figure 4.3 Chemical structure of α-tocopherol ($C_{29}H_{46}O_2$).

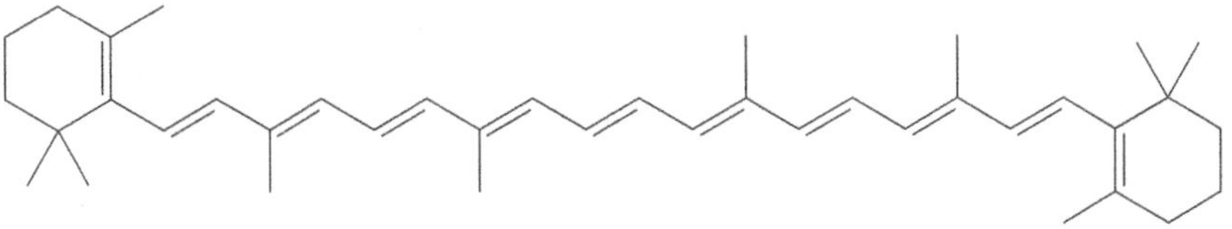

Figure 4.4 Chemical structure of β-carotene ($C_{40}H_{56}$).

shows that β-carotene is the major component. Figure 4.4 shows the chemical structure of β-carotene.

4.5 Development of Aspen Plus® Physical Property Database for Palm Oil Components

The development of a palm oil physical property database is a prerequisite for simulation modeling of SFE processes using Aspen Plus® process simulator. There are a number of important pure component physical properties required to accurately model the solubility of a compound in supercritical fluids using the equation of state model. Physical property data for the components involved in the thermodynamic modeling of CPO-supercritical CO_2 systems were not available in the Aspen Plus® property database, even though some physical property data for the individual components in the palm oil mixture were available in open literature. However, many of the physical properties of palm oil components cannot be determined experimentally due to thermal decomposition of the components at temperatures below their boiling points. It was therefore necessary to estimate the critical properties of these components theoretically. These properties can actually be considered as hypothetical properties. Once the critical properties had been estimated, a set of physical properties for the components of interest could be determined using the built-in Property Constant Estimation System in Aspen Plus® process simulator.

4.5.1 Vapor Pressure Estimation

Accurate vapor pressure data are required for the development of reliable thermodynamic models such as equations of state (Ashour 1989). The vapor pressure of the solute, which is a strong function of system temperature, is an important physical property in the equation of state model for solubility in supercritical fluids. However, measurements of vapor pressure and boiling point data are not easily determined for palm oil components such as triglycerides due to its complex nature (chemical composition) and temperature sensitivity.

Extrapolative and predictive methods were used to estimate the vapor pressures for fatty oil components due to the difficulties in carrying out vapor pressure measurements at low temperatures. Equation 4.10 represents the extended Antoine model (Aspen Tech 2000) used in this study to extrapolate the available experimental data to the temperature of interest:

$$\ln P^{sat} = A + \frac{B}{T} + CT + D \ln T + E \ln T^6 \qquad (4.10)$$

where A, B, C, D, and E are the Antoine equation parameters.

The parameters of the Antoine equation were obtained by minimizing the objective function, Q, using the generalized least squares regression. The least squares algorithm is available in Aspen Plus® release 10.2.1.

$$\text{Objective function, } Q = \sum_{i=1}^{m} \left[\frac{P_i^{calc} - P_i^{exp}}{P_i^{exp}} \right]^2 \qquad (4.11)$$

where

m = number of experimental data points
P_i^{calc} = calculated vapor pressure
P_i^{exp} = experimental vapor pressure

4.5.2 Estimation of Pure Component Critical Properties

Palm oil components decompose before reaching their normal boiling temperature and therefore the critical properties required by the equation of state have to be estimated. The method proposed by Dohrn and Brunner (1994) was used, as it required only the liquid molar volume at 20 °C, $V_{L,20}$, and vapor pressure data as the input information to obtain the critical properties of palm oil components.

4.5.2.1 Critical Properties Estimation Using Normal Boiling Point

For two-parameter equation of state, the pure component parameter a_i and b_i for the critical temperature can be determined from the conditions at the critical point:

$$\frac{a_i}{\Omega_a} = \frac{R^2 T_{ci}^2}{P_{ci}} \qquad (4.12)$$

$$\frac{b_i}{\Omega_b} = \frac{RT_{ci}}{P_{ci}} \qquad (4.13)$$

where Ω_a and Ω_b are equation specific constants, i.e. for the Peng–Robinson equation of state (Peng and Robinson 1976), $\Omega_a = 0.457\,24$ and $\Omega_b = 0.0778$.

Dohrn and Brunner (1994) proposed to calculate a_i and b_i from the liquid molar volume $V_{L,20}$ and T_{bi} as follows:

$$\frac{b_i}{\Omega_b} = b^{(1)} v_{L,20} T_{bi} + b^{(2)} \qquad (4.14)$$

$$\frac{a_i}{\Omega_a} = a^{(1)} \left(\frac{b_i}{\Omega_b} T_{bi} \right)^{a^{(2)}} \qquad (4.15)$$

Coefficients $b^{(1)}$ and $b^{(2)}$ have been determined by correlating the data of 380 fluids. For the all-fluid correlation, $a^{(1)} = 21.269\,24$ kJ/kmol/K, $a^{(2)} = 0.913\,049$, $b^{(1)} = 0.025\,561\,88$ K^{-1}, and $b^{(2)} = 0.168\,721$ m^3/kmol.

When a_i and b_i are known, T_{ci} and P_{ci} can be determined using Eqs. 4.12 and 4.13:

$$T_{ci} = \frac{a_i}{\Omega_a} \frac{\Omega_b}{b_i} \frac{1}{R} \qquad (4.16)$$

$$P_{ci} = \frac{a_i}{\Omega_a} \left(\frac{\Omega_b}{b_i} \right)^2 \qquad (4.17)$$

Pitzer method (Reid et al. 1987) was used to find a correlation for the acentric factor ω. The acentric factor is an empirical correction introduced to extend cubic EOS to slightly nonspherical molecules as long molecules; it accounts for the shape of the molecule and is 0 when the molecule is spherical:

$$\omega = -\log_{10}(P_r^{\mathrm{sat}})_{T_r = 0.7} - 1.0 \qquad (4.18)$$

with P_r^{sat} being the reduced vapor pressure. Assuming that $\log_{10}(P_r^{\mathrm{sat}})$ is a linear function of $1/T_r$:

$$\log_{10} P_r^{\mathrm{sat}} = c\left(1 - \frac{1}{T_r}\right) \qquad (4.19)$$

where c is a fluid specific constant, which can be determined using the normal boiling point:

$$c = \frac{\log_{10}(101.3\text{ kPa}/P_c)}{1 - T_c/T_b} \qquad (4.20)$$

Replacing $\log_{10}(P_r^{sat})$ in Eq. 4.19 with Eqs. 4.18 and 4.20 leads to

$$\omega = -\frac{3}{7}\frac{\log_{10}(101.3\,\text{kPa}/P_c)}{(T_c/T_b - 1)} - 1 \tag{4.21}$$

4.5.2.2 Critical Properties Estimation Using One Vapor Pressure Point

This method can be used if T_b is not known, but with one experimental vapor pressure point for an arbitrary temperature T_1 is available. As input information $V_{L,20}$, $P^{sat}(T_1)$ and initial value for T_c and P_c (e.g. by using Joback's group contribution method) are needed. The fluid-specific vapor pressure constant c can be found by

$$c = \frac{\log_{10}(P^{sat}(T_1)/P_c)}{1 - T_c/T_1} \tag{4.22}$$

Initial value for T_b can then be calculated by solving Eq. 4.20 for T_b:

$$T_b = \frac{T_c}{\left(1 - \frac{\log_{10}(101.3\,\text{kPa}/P_c)}{c}\right)} \tag{4.23}$$

a_i and b_i can be calculated from $V_{L,20}$ and T_b using the all-fluid correlations in Eqs. 4.14 and 4.15. With Eqs. 4.16 and 4.17, new value for T_c^{new} and P_c^{new} can be determined. A new iteration is started until T_c and P_c converge within tolerance (i.e. $\varepsilon = 10^{-8}$). Figure 4.5 shows the logic diagram for the calculation of T_c, P_c, and T_b. The iterative calculations using the above methods had been performed with the aid of the spreadsheet program, Microsoft® Excel.

4.6 Binary Interaction Parameters Calculations

For the prediction of phase equilibrium of palm oil-supercritical CO_2 system, binary interaction parameters for EOS are needed for dissimilar components. The binary interaction parameters are used to improve the prediction of the EOS. To calculate the binary interaction parameters for the RKA-EOS, the interaction parameters were initially assumed to be zero. The optimized binary interaction parameters were found by performing a maximum likelihood (errors-in-variables) estimation using Deming algorithm (Aspen Tech 2000) to minimize the following objective function:

$$Q = \sum_N \left(\frac{T^{exp} - T^{calc}}{\sigma_T}\right)^2 + \sum_N \left(\frac{P^{exp} - P^{calc}}{\sigma_P}\right)^2 + \sum_N \left(\frac{x_i^{exp} - x_i^{calc}}{\sigma_{x_1}}\right)^2 + \sum_N \left(\frac{y_i^{exp} - y_i^{calc}}{\sigma_{y_1}}\right)^2$$

$$\tag{4.24}$$

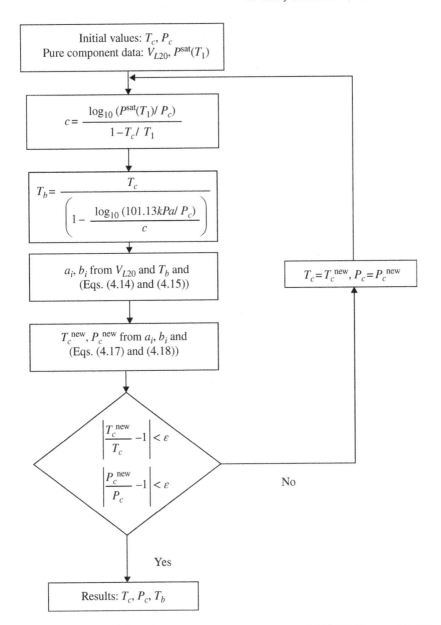

Figure 4.5 Calculations of T_c, P_c, and T_b from $V_{L,20}$ and $P^{sat}(T_1)$ (*Source:* Modified from Dohrn and Brunner (1994). © John Wiley & Sons).

where

T, P = Temperature and pressure of the studied data
x_i, y_i = Composition of a component i in liquid and fluid phase
σ = Standard deviation
calc, exp = Denotes calculated and experimental values
N = Denotes number of data points

The quality of the regressed data was assessed using the average absolute deviation (AAD) between the measured and calculated data:

$$\text{AAD}(\%) = \frac{1}{N}\sum_{i=1}^{N}|d_i| \times 100\% \qquad (4.25)$$

where

d_i = Error between the experimental and calculated values
N = Number of data points

Phase equilibrium data available in the literature for the binary system of palm oil components-supercritical CO_2 are presented in Table 4.2. The

Table 4.2 Phase equilibrium data of palm oil related component-CO_2 system.

Palm oil component-CO_2	Temperature (K)	Pressure (MPa)	References
Tripalmitin	333	20–50	Weber and Brunner (1995)
	353		Weber et al. (1999)
Triolein	313	15–30	Bharath et al. (1993)
	333	20–50	Weber et al. (1999)
	353	10–50	Weber and Brunner (1995)
Oleic acid	313	6–31	Yu et al. (1992)
	333, 353	15–30	Bharath et al. (1992)
α-Tocopherol	313, 333	10–24	Chen et al. (2000)
	323	9–26	Pereira et al. (1993)
	343	26–35	Meier et al. (1994)
β-Carotene	313, 323, 333	20–28	Sovová et al. (2001)
	353	20–32	Johannsen and Brunner (1997)

solubility of β-carotene in supercritical CO_2 was calculated by assuming a liquid-fluid equilibrium using the RKA thermodynamic model. β-Carotene (solute) was treated as a "liquid" component since Aspen Plus® does not generally deal with the solid–liquid–fluid equilibrium (Aspen Tech 2000).

In this study, the solute–solute interaction parameters between palm oil components (TG, FFA, and minor components) were assumed to be zero. The reason behind this was that experimental data for the TG-FFA (for palm oil) was not available; hence, binary interaction parameters between these components could not be estimated. Since the two components are very dissimilar. The approximation might not be completely adequate. Nevertheless, satisfactory results have been reported in the literature from predicting the phase equilibrium of a complex oil mixture based on only supercritical solvent–solute interactions while neglecting the solute–solute interactions. Some examples include the prediction of phase equilibrium for a CO_2 – soybean oil deodorizer condensates system by Araújo et al. (2001), CO_2 – citrus peel oil system by Espinosa et al. (2000), and CO_2 – essential oil by Sovová et al. (2001).

4.7 Supercritical Fluid Extraction Process Development

Multistage supercritical fluid extraction has emerged as an alternative to replace traditional separation processes, when the separation of thermally labile substances and the attainment of high purity products are the targets (Catchpole et al. 2000). Economically, a countercurrent SFE process is more advantageous since the saturation solubility of supercritical fluid can be maintained.

4.7.1 Hydrodynamics of Countercurrent SFE Process

The viability of packed column fractionation of lipid mixtures, including palm oil, depends on there being sufficient density difference between the solvent and solute phases to avoid entrainment and flooding (Tegetmeier et al. 2000). The study of the fluid dynamic behavior of the palm oil-supercritical CO_2 system determines the range of temperature and pressure of which countercurrent extraction is feasible.

The density difference of coexisting phases is crucial as a limiting factor to assure countercurrent flows and it must always be higher than 150 kg/m^3 in order to avoid flooding (Gast et al. 2001). For palm oil-supercritical CO_2 systems, the density of coexisting phases of the CPO-CO_2 system was measured by Kalra et al. (1987), Machado and Brunner (1997), and Tegetmeier et al. (2000). The results of these measurements are presented in Figure 4.6. The figure shows that density of palm oil in supercritical CO_2 increases with increasing pressure and decreasing temperature. The pressure dependence of the liquid density of palm oil in contact with supercritical CO_2 seems to be linear (Tegetmeier et al. 2000). The incorporation of CO_2 into the palm oil increases the density of the liquid phase, although the density of pure CO_2 is always lower than the density of palm oil at the temperatures and pressures of interest.

Figure 4.7 shows the density differences between the coexisting CO_2-rich and oil-rich phases at temperatures between 333 and 373 K. As shown in the figure, the required density difference at a temperature of 333 K is not complied above pressure of 22 MPa which is within the pressure range of interest. Thus, only a temperature range of 353–373 K and pressure range of 20–30 MPa which gave sufficiently high-density difference were considered for the countercurrent SFE process operation.

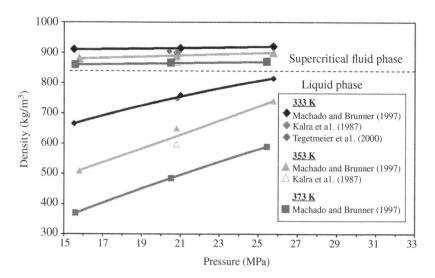

Figure 4.6 Density of coexisting phases for palm oil-supercritical CO_2 system.

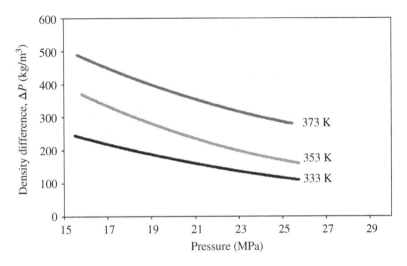

Figure 4.7 Density difference between coexisting phases for palm oil-CO_2 system.

4.7.2 Solubility of Palm Oil in Supercritical CO_2

Solubility of certain compounds in the supercritical phase is the most important parameter for supercritical extraction (Brunner 1998). Solubility behavior of oil triglycerides in supercritical CO_2 as a function of pressure, temperature, and solvent density has been previously discussed (Tilly et al. 1990).

Solubility is used to define the maximum yield obtainable in an extraction process occurring at hypothetically infinite contact time. Therefore, the oil concentration in the supercritical phase is lower than the equilibrium value because thermodynamic equilibrium cannot be reached in an industrial extractor owing to the finite contact time between the solvent and the solute (Markom et al. 1999). Reverchon and Osseo (1994) and Stahl et al. (1986) suggest an extraction efficiency of 60% with respect to the equilibrium value in their work involving soybean oil-supercritical CO_2 systems. Ooi et al. (1996) observed that the solubility of palm oil in the equilibrium state is higher than that under continuous processing conditions. Therefore, the empirical extraction stage efficiency, η_E, must be used to correct for the departures of oil solubility under continuous operating conditions from that of the equilibrium solubility:

$$\eta_E = \frac{\text{Solubility (continuous processing)}}{\text{Solubility (equilibrium condition)}} \times 100\% \quad (4.26)$$

Pietsch and Eggers (1999) have obtained the experimental efficiencies of separation for CO_2-water systems. In their study, the efficiency of separation is found to decrease with increasing S/F ratio. In this study, extraction efficiency was defined as a function of solvent-to-feed ratio (S/F). Figure 4.8 shows the computed extraction efficiency based on the experimental solubility data of Rui Ruivo et al. (2001) for edible oil-CO_2 system. An increase in the S/F ratio (higher gas flow rates and/or lower liquid feed flow rates) resulted in a decrease in the extract phase loading due to the fact that as S/F increases lower mass of liquid would be available for complete saturation of fluid phase (Rui Ruivo et al. 2001).

4.7.3 Process Modeling and Simulation

The design of countercurrent multistage SFE process requires rigorous calculation of stream flow rates, stream compositions, concentration profiles along the column, temperatures, and pressures at each stage. Commercial

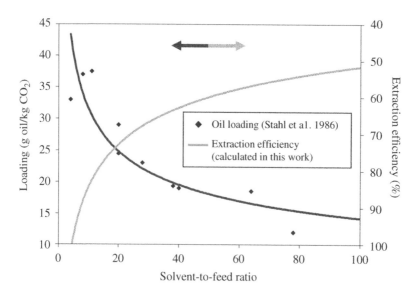

Figure 4.8 Loading of oil in CO_2 (*Source:* Stahl et al. (1986). © Springer Nature) and computed extraction efficiency as a function of S/F ratio.

process simulation software, Aspen Plus® release 10.2.1 was used to solve flow sheet modeling problems encountered in this work to obtain a better insight into compositions of phases along the separation process. Aspen Plus® process simulator includes thermodynamic and physical properties, unit operations, hydrodynamics, cost analysis, optimization, which makes the process simulator a powerful numerical tool to perform complete process design analysis. Simulations were performed on a stand-alone PC with a Microsoft® Windows operating system running with a Pentium III/1.0 GHz processor.

The calculations for the countercurrent SFE process were performed using the concept of theoretical stages in which the extractor was split into cascades of flash modules. Each stage of the column was assumed as a single flash at fixed temperature and pressure. Using the modular structure of the Aspen Plus®, a rigorous sequential simulation for the process had been implemented using a flash separator (FLASH2) unit to model the cascade of flash modules in accordance with theory of a theoretical separation unit. The process flow diagram of the supercritical CO_2 extraction column is shown in Figure 4.9.

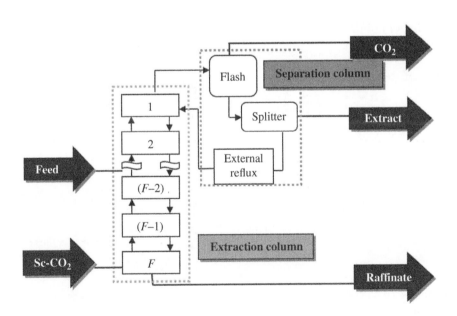

Figure 4.9 Process flow diagram for SFE of palm oil using CO_2.

A variation of the number of theoretical stages and other process variables provides information on the purity of the products with different number of theoretical stages and operating conditions.

4.7.3.1 Simple Countercurrent Extraction

In a simple countercurrent extraction scheme, the column works as a stripping section, where supercritical CO_2 is the continuous phase entering at the bottom of the column and feed oil is the dispersed phase entering at the top of the extraction column. Figure 4.10 shows the basic flow scheme of a simple countercurrent extraction scheme with low product recovery.

4.7.3.2 Countercurrent Extraction with External Reflux

For large-scale continuous SFE processes, an external reflux is a more effective way to increase process efficiency (Espinosa et al. 2000). The reflux is needed to create a countercurrent flow regime inside the column and the reflux flow should be kept at a minimum for economic reasons. The method requires that sufficient solvent be removed from the extract leaving the cascade to form a raffinate, part of which is returned to the cascade as reflux, the remainder being withdrawn from the plant as a product. Raffinate is withdrawn from the cascade as a bottoms product, and fresh solvent is admitted directly to the bottom of the cascade.

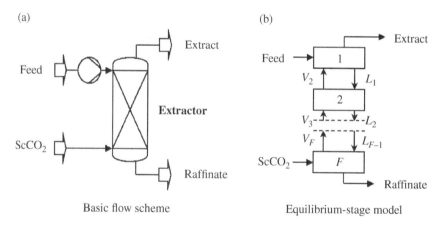

Figure 4.10 Schematic diagram for simple countercurrent SFE process scheme.

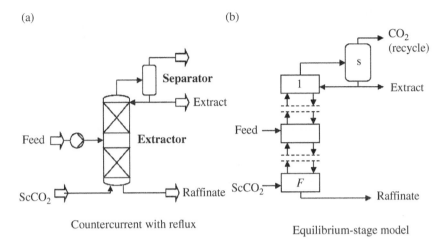

Figure 4.11 Schematic diagram for countercurrent (with reflux) SFE process scheme.

In this scheme, the extract is depressurized, and part of the separator liquid is recycled to the extractor as reflux. The separator connected to the top of the extractor is represented by the FLASH2 module in Aspen Plus®. The existence of reflux flow increases the liquid flow rate in the column and with the consequent increase in raffinate (refined palm oil) recovery. Figure 4.11 shows the flow scheme of a countercurrent extraction column with reflux.

4.7.4 Process Analysis and Optimization

The variables that have the greatest impact on the SFE process are the extraction column pressure and temperature, the number of stages in the extraction column, S/F ratio and the reflux ratio. In this work, the deacidification of CPO with supercritical CO_2 was evaluated. Different extraction schemes were studied, and optimal operating conditions were determined. SFE process schemes considered for detailed study were (i) simple countercurrent extraction; (ii) countercurrent extraction with external reflux. Two objective functions were considered: (i) maximum recovery of refined palm oil and, (ii) minimum CO_2 solvent recirculation.

Part II: Results and Discussion

To establish commercial SFE processes for palm oil processing, it is imperative to have reliable phase equilibrium data. The RKA thermodynamic model was used to correlate and predict the phase equilibrium for the CPO-supercritical CO_2 system. The key steps in thermodynamic modeling involved the characterization of palm oil mixture, the estimation of pure component vapor pressures and critical properties, and the regression of experimental phase equilibrium data for the palm oil component-supercritical CO_2 binary system available in the literature to yield the binary interaction parameters for the RKA-EOS. The results involved in modeling the phase equilibrium of palm oil components with supercritical CO_2 are presented in this chapter. The ultimate aim of modeling was to generate reliable solubility data and distribution coefficients for palm oil components in supercritical CO_2, which are crucial for process design and optimization of a separation system using the SFE technique.

4.8 Palm Oil Component Physical Properties

Pure component physical properties including the vapor pressure, critical temperature, and critical pressure of palm oil components are required to generate the phase equilibrium data in the Aspen Plus®. Critical temperature and critical pressure for palm oil components cannot be measured experimentally because these components thermally degrade at temperatures far below their critical points. Thus, the critical properties for palm oil components have to be estimated. The estimated physical properties for palm oil components are given in the following section.

4.8.1 Vapor Pressure of Palm Oil Components

Only limited vapor pressure data could be found for tripalmitin, triolein, and α-tocopherol in the open literature (Mag 1994; Perry et al. 1949), and they covered a temperature range considerably above the ones of interest in this study. Perry et al. (1949) studied the vapor pressures of saturated (i.e. tripalmitin) and unsaturated TG as well as mixed TG (i.e. refined soybean oil and olive oil). The vegetable oils were found to present similar vapor pressure curves. Vapor pressure of triolein is regarded as similar to the vapor pressure of olive oil due to the fact that the principal fatty acid constituent of this olive oil is oleic acid, which makes it very similar to triolein (Perry et al. 1949).

Vapor pressure for unsaturated oleic acid is not available in the literature. Therefore, vapor pressure data of stearic acid, which has the same chain length but a different degree of unsaturation, was used in this study. Perry et al. (1949) found that the effect of unsaturation has very little influence (difference by few degrees Celsius) on the vapor pressure of triglyceride and methyl ester of fatty acids. For β-carotene, the sublimation pressure reported by Cygnarowicz et al. (1990) was used.

The extended Antoine model (see Eq. 4.10) was used to correlate and extrapolate the available experimental vapor pressure data for palm oil components to the temperatures of interest. The Mani's method (Aspen Tech 2000) was used to estimate parameters for the extended Antoine vapor pressure equation for complex palm oil components that decompose at temperatures below the normal boiling point. The correlated parameters for the extended Antoine vapor pressure equation are summarized in Table 4.3.

The extended Antoine vapor pressure model was used to correlate the experimental data reported in the literature. Figure 4.12 shows the relative deviations of the experimental and calculated palm oil components vapor pressure at various temperatures. The figure indicates that the relative deviations between the correlated vapor pressure and the experimental vapor pressure data reported in the literature are randomly distributed. Thus, the vapor pressure model used in this study can adequately correlate the vapor pressure data of palm oil components (Shacham et al. 1995).

Table 4.3 Pure component parameters for the extended Antoine equation.

Component	A	B	C	D	E
Tripalmitin	212.57	−32 864	—	−25.22	1.34×10^{-18}
Triolein	222.89	−34 577	—	−26.49	1.33×10^{-18}
Oleic acid	−577.87	7558	−0.134	101.47	1.77×10^{-17}
α-tocopherol	119.91	−19 461	—	−13.59	8.53×10^{-19}
β-carotene	72.13	−13 969	—	−7.56	3.13×10^{-19}

Source: Aspen Tech (2000). © Aspen Technology, Inc.

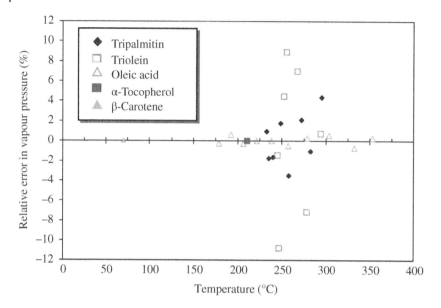

Figure 4.12 Relative deviation of calculated vapor pressure from experimental vapor pressure of palm oil components.

4.8.2 Pure Component Critical Properties

The method proposed by Dohrn and Brunner (1994) allows the estimation of pure component critical properties from liquid molar volume at 20 °C and vapor pressure data as input information. Table 4.4 summarizes the predicted physical properties of palm oil components investigated in this work.

4.9 Regression of Interaction Parameters for the Palm Oil Components-Supercritical CO_2 Binary System

Owing to the extremely low-vapor pressures of triglycerides and the considerable difficulty in predicting the vapor pressure of triglycerides from experiments carried out only at high temperatures, the polar factor for palm oil components was computed simultaneously with binary interaction from palm oil component-supercritical CO_2 phase equilibrium data.

Table 4.4 Predicted physical properties of pure components in palm oil.

Component	T_b (K)[a]	T_c (K)[a]	P_c(kPa)[a]	ω[b]	$V_{L,20}$[c] (m³/kmol)
Tripalmitin	864.21	947.10	396.82	1.6500	0.8906
Triolein	879.92	954.10	360.15	1.8004	0.9717
Oleic acid	646.52	813.56	1250.19	0.8104	0.3172
α-Tocopherol	794.52	936.93	838.45	1.1946	0.4533
β-Carotene	908.58	1031.06	678.41	1.6255	0.5348

[a] Estimated by the method of Dohrn and Brunner (1994).
[b] Estimated using the Pitzer method (Reid et al. 1987).
[c] Liquid molar volume data (at 20 °C) obtained from open literature: tripalmitin, triolein, oleic acid (Formo et al. 1979); α-tocopherol (Dohrn and Brunner 1994); β-carotene (Weast et al. 1990).
Sources: Dohrn and Brunner (1994), Reid et al. (1987), Formo et al. (1979), and Weast et al. (1990). © John Wiley & Sons.

4.9.1 Binary System: Triglyceride – Supercritical CO_2

The correlation of phase equilibrium data of tripalmitin-CO_2 and triolein-CO_2 was carried out for the literature data reported by Weber et al. (1999) and Weber and Brunner (1995) at a temperature range of 333–353 K. The data published by various authors are shown in Figures 4.13–4.16. It is well known that differences can occur as a result of different experimental setups or differences in the purity of the feed materials (Üstündağ and Temelli 2000). In this study, only the data from Weber et al. (1999) and Weber and Brunner (1995) were considered in the correlation in order to maintain consistency.

Tables 4.5 and 4.6 show the values of the calculated polar factor (η), binary interaction parameters (k_a, k_b), and the absolute average deviation (AAD) for the tripalmitin-supercritical CO_2 system and the triolein-supercritical CO_2 system.

The calculated liquid and supercritical fluid phase composition for the binary system triglyceride-supercritical CO_2 system are shown in Figures 4.13 and 4.14 (for tripalmitin-CO_2 system), and Figures 4.15 and 4.16 (for triolein-CO_2 system), respectively.

For triglycerides-CO_2 system, the vapor-liquid equilibrium exhibits open-ended top saturation loop in which they remain open at the highest pressure explored (50 MPa). It was also found that a considerable amount of CO_2

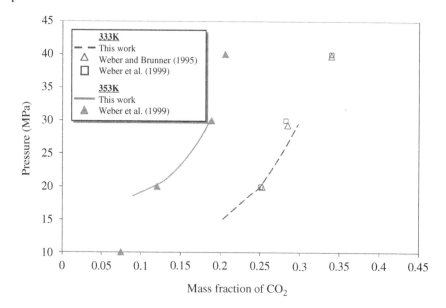

Figure 4.13 Liquid phase compositions for the tripalmitin – supercritical CO_2 system at 333–353 K.

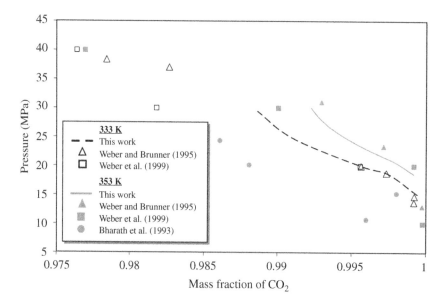

Figure 4.14 Fluid phase compositions for the tripalmitin – supercritical CO_2 system at 333–353 K.

4.9 Regression of Interaction Parameters

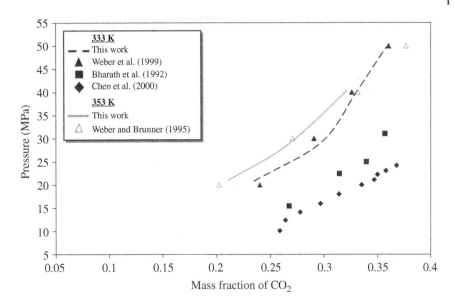

Figure 4.15 Liquid phase compositions for the triolein – supercritical CO_2 system at 333–353 K.

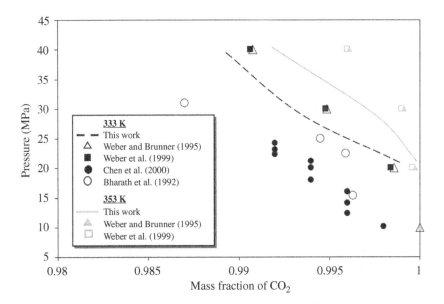

Figure 4.16 Fluid phase compositions for the triolein – supercritical CO_2 system at 333–353 K.

Table 4.5 Optimal parameters for the RKA model for the tripalmitin (1) – supercritical CO_2 (2) binary system.

T (K)	η_1	$k_{a,12}$	$k_{b,12}$	AAD$_x$ (%)	AAD$_y$ (%)
333	−2.3962	0.0396	−0.0186	1.4700	0.0629
353	−1.8232	0.0547	−0.0318	0.8156	0.0780

Table 4.6 Optimal parameters for the RKA model for the triolein (1)– supercritical CO_2 (2) binary system.

T (K)	η_1	$k_{a,12}$	$k_{b,12}$	AAD$_x$ (%)	AAD$_y$ (%)
333	−3.5957	0.0289	−0.0154	1.5689	0.3559
353	−3.2240	0.0392	−0.0065	2.9523	0.1929

dissolves into the liquid phase for both tripalmitin-CO_2 and triolein-CO_2 systems. The solubilities of carbon dioxide in tripalmitin and triolein increase with decreasing temperature (see Figures 4.13 and 4.15).

Figures 4.14 and 4.16 show that the solubilities of tripalmitin and triolein in supercritical CO_2 increase with pressure. The solubilities of tripalmitin and triolein in supercritical CO_2 show similar behavior: the solubility increases with decreasing temperature. The amount of oil dissolved in the supercritical CO_2 phase increases with the increase in pressure. The solubility of tripalmitin in supercritical CO_2 is generally larger than that of triolein. This is due to the difference in molecular size of these triglycerides. Higher molecular weight triglycerides are less soluble in supercritical CO_2.

4.9.2 Binary System: Oleic Acid – Supercritical CO_2

The correlation of phase equilibrium data of oleic acid-supercritical CO_2 reported by Yu et al. (1992) and Bharath et al. (1992) was carried out. Table 4.7 shows the calculated pure component polar factor (η), binary interaction parameters (k_a, k_b), and the absolute average deviation (AAD) for the oleic acid – supercritical CO_2 system. The AAD values for the liquid phase and supercritical fluid phase compositions indicate a good agreement with

Table 4.7 Optimal parameters for the RKA model for the oleic acid (1) – supercritical CO_2 (2) binary system.

T (K)	η_1	$k_{a,12}$	$k_{b,12}$	AAD$_x$ (%)	AAD$_y$ (%)
313	−1.2127	0.0736	−0.0027	0.6222	0.0455
333	−1.0873	0.0819	0.0036	0.9410	0.4304
353	−1.0366	0.0910	0.0141	0.5377	0.1025

experimental data. The overall AAD calculated for this system is 0.4466% which is within the experimental measurement error of 5% (max) in the work of (Bharath 1993).

The calculated liquid and supercritical fluid phase composition for the oleic acid – supercritical CO_2 system at a temperature range of 313–353 K are shown in Figures 4.17 and 4.18, respectively. The data published by various authors are also shown in the figures. For oleic acid-CO_2 systems, the vapor-liquid equilibrium exhibits open-ended top saturation loop in which they remain

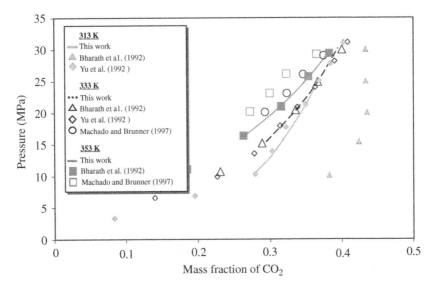

Figure 4.17 Liquid phase compositions for the oleic acid – supercritical CO_2 system at 313–353 K.

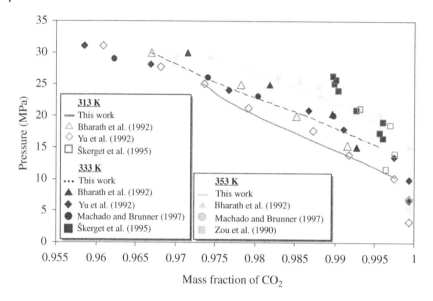

Figure 4.18 Fluid phase compositions for the oleic acid – supercritical CO_2 system at 313–353 K.

open up to the highest pressure explored (30 MPa). It was also found that a considerable amount of CO_2 dissolves into the liquid phase for oleic acid-CO_2 systems. Thus, the knowledge of the liquid phase composition is very important in modeling and characterizing the palm oil-supercritical CO_2 system. Figure 4.18 shows that the solubility of oleic acid increases with pressure. Comparisons among Figures 4.14, 4.16, and 4.18 show that solubility of oleic acid is higher than that of triglycerides (tripalmitin and triolein) at pressure range of between 20 and 30 MPa. This indicates that fatty acids and triglycerides can be easily separated using supercritical carbon dioxide.

4.9.3 Binary System: α-Tocopherol – Supercritical CO_2

Correlations of the phase equilibrium data of Chen et al. (2000), Meier et al. (1994), and Pereira et al. (1993) were carried out. Table 4.8 shows the values of the calculated polar factor (η), binary interaction parameters (k_a, k_b), and the absolute average deviation (AAD) for the α-tocopherol – supercritical CO_2 system. The AAD values for the liquid phase and supercritical fluid phase compositions are 1.35 and 0.18%, respectively. The low AAD values indicate a good agreement between correlated values and the experimental data.

Table 4.8 Optimal parameters for the RKA model for the α-tocopherol (1) – supercritical CO_2 (2) binary system.

T (K)	η_1	$k_{a,12}$	$k_{b,12}$	AAD_x (%)	AAD_y (%)
313	−1.1016	0.0479	−0.0309	0.0606	0.0279
323	−0.8064	0.0568	−0.0206	2.7648	0.1693
333	−0.5233	0.0630	−0.0261	1.0116	0.1726
343	0.1773	0.0927	0.0012	0.4509	0.0751
353	0.3609	0.1030	0.0599	2.4576	0.4649

The calculated liquid and supercritical fluid phase composition for the α-tocopherol-supercritical CO_2 system at a temperature range of 313–353 K are presented in Figures 4.19 and 4.20. The data published by various authors are also shown in the figures.

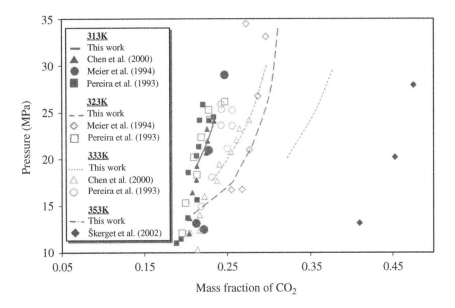

Figure 4.19 Liquid phase compositions for the α-tocopherol – supercritical CO_2 system at 313–353 K.

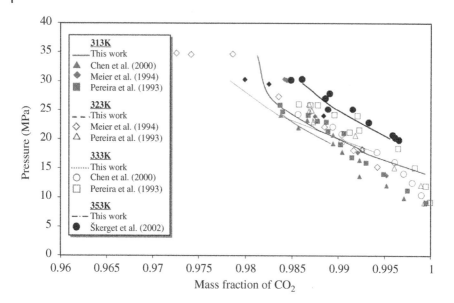

Figure 4.20 Fluid phase compositions for the α-tocopherol – supercritical CO_2 system at 313–353 K.

For α-tocopherol-supercritical CO_2 systems, it was found that a considerable amount of CO_2 dissolves into the liquid phase for tocopherol-CO_2 systems. Figure 4.20 shows that the solubility of tocopherol increases with pressure.

4.9.4 Binary System: β-Carotene – Supercritical CO_2

Correlations of the phase equilibrium data of Johannsen and Brunner (1997) and Sovová et al. (2001) were carried out. Table 4.9 shows the calculated polar factor (η), binary interaction parameters (k_a, k_b), and the absolute average

Table 4.9 Optimal parameters for the RKA model for the β-carotene (1) – supercritical CO_2 (2) binary system.

T (K)	η_1	$k_{a,12}$	$k_{b,12}$	AAD$_y$ (%)
313	−1.5747	0.0421	−0.1107	6.8471
323	−1.3790	0.0458	−0.1148	3.3289
333	−1.0863	0.0545	−0.1194	9.3478
353	−0.7619	0.0584	−0.1310	1.5822

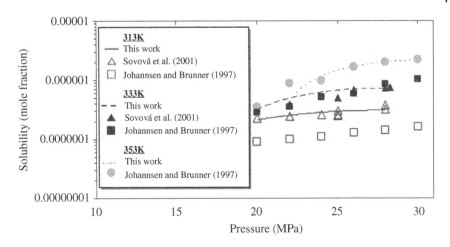

Figure 4.21 Solubility of β-carotene in supercritical CO_2.

deviation (AAD) for the β-carotene – supercritical CO_2 system. The calculated solubilities of β-carotene in supercritical CO_2 and the literature solubility data are shown in Figure 4.21. The figure presents the β-carotene solubility behavior in supercritical CO_2 as a function of pressure. The overall AAD for supercritical CO_2 phase composition is 5.28%. In view of the low solubility of carotene in supercritical CO_2, typically in the order of 10^{-7} mol fraction, the correlation is considered to be acceptable.

4.9.5 Temperature-Dependent Interaction Parameters

The values of polar factor (η) and binary interaction parameters ($k_{a,12}$ and $k_{b,12}$) at temperatures between 313 and 353 K were obtained by nonlinear regression procedure applied to the sets of experimental data for palm oil components-supercritical CO_2 systems in the previous section. The calculations demonstrated that the phase equilibria of the palm oil components-CO_2 systems was correlated very well with RKA-EOS with a conventional quadratic mixing rule for the mixture cross energy and covolume parameters. The maximum AAD of the RKA-EOS predictions from these data were calculated to be 2.95% for the liquid phase and 0.46% for the fluid phase.

The temperature-dependent polar factors and binary interaction parameters were then obtained by correlating the polar factor and binary interaction parameters using the linear regression function in Microsoft® Excel. The resulting temperature-dependent polar factors and binary interaction parameters (with

Table 4.10 Temperature-dependent polar factors and binary interaction parameters for palm oil components-supercritical CO_2 systems.

	Polar factor (η)	Binary interaction parameters
Tripalmitin-CO_2	$0.0281\ T - 11.765$	$k_a = 0.0007\ T - 0.2117$
		$k_b = -0.0006\ T + 0.2021$
Triolein-CO_2	$0.0186\ T - 9.7848$	$k_a = 0.0005\ T - 0.1428$
		$k_b = 0.0004\ T - 0.1632$
Oleic acid-CO_2	$0.0044\ T - 2.5779$	$k_a = 0.0004\ T - 0.0622$
		$k_b = 0.0004\ T - 0.1349$
α-Tocopherol-CO_2	$0.0427\ T - 14.6165$	$k_a = 0.0017\ T - 0.5082$
		$k_b = 0.0059\ T - 2.0272$
β-Carotene-CO_2	$0.0206\ T - 8.0155$	$k_a = 0.0004\ T + 0.0895$
		$k_b = -0.0005\ T + 0.0501$

$R^2 > 0.9$) are shown in Table 4.10. These values were used to predict the phase equilibrium of palm oil components-CO_2 system for temperature range of 313–373 K and pressure range of 20–30 MPa.

4.10 Phase Equilibrium Calculation for the Palm Oil – Supercritical CO_2 System

Palm oil was approximated as a mixture of simple triglycerides, namely tripalmitin (C48) and triolein (C54), and other minor components. In order to validate this assumption, several trial calculations were made using the $P–T$ isothermal flash algorithm implemented in Aspen Plus®.

Using the calculated polar factor for palm oil components and the binary interaction parameters for the palm oil component-supercritical CO_2, the phase equilibrium calculations were carried out using a $P–T$ isothermal flash algorithm implemented in Aspen Plus®. Phase equilibrium (flash) calculations were performed throughout a multicomponent process simulation run for a supercritical fluid-liquid split to obtain the phase compositions of palm oil components.

Table 4.11 provides a comparison between phase equilibria of palm oil-supercritical CO_2 systems reported in the literature (Brunner 1978; Kalra

Table 4.11 Comparison of predicted phase equilibrium for a palm oil – CO_2 system with literature data[a].

Temperature (°C)	Pressure (bar)	CO_2 in liquid (wt%), x_{CO_2}			CO_2 in fluid (wt%), y_{CO_2}		
		This work	Exp. data	Deviation[b] (%)	This work	Exp. data	Deviation[c] (%)
50	208.1	30.23	38.6[i]	21.68	99.26	98.61[i]	0.66
60	208.2	25.92	36.3[i]	28.60	99.53	99.24[i]	0.29
80	208.0	19.81	33.1[i]	40.15	99.78	99.75[i]	0.03
70	200.0	21.65	23.66[ii]	8.50	99.75	99.72[ii]	0.03
70	300.0	28.59	27.64[ii]	3.44	99.02	98.91[ii]	0.11
70	350.0	29.87	29.73[ii]	0.47	98.88	98.16[ii]	0.73
75	202.0	20.44	25.3[iii]	19.21	99.78	99.79[iii]	0.01
75	308.0	27.72	31.7[iii]	12.56	99.04	98.83[iii]	0.21

[a] Experimental data by (i) Kalra et al. (1987); (ii) Stoldt and Brunner (1998); (iii) Brunner (1978).
[b] For liquid phase: Deviation = $|(x^{exp} - x^{calc})|/x^{exp} \times 100\%$;
[c] For supercritical fluid phase: Deviation = $|(y^{exp} - y^{calc})|/y^{exp} \times 100\%$.
Sources: Kalra et al. (1987), Stoldt and Brunner (1998), and Brunner (1978). © John Wiley & Sons.

et al. 1987; Stoldt and Brunner 1998) and that predicted by the RKA model in this work. Results were obtained within 50 iterations corresponding to computation times of less than 0.5 CPU seconds. The phase equilibrium of palm oil-supercritical CO_2 predicted using the RKA-EOS agrees well with the experimental data of Stoldt and Brunner (1998) and Brunner (1978). The overall AAD of the RKA-model predictions were calculated to be 16.83% for the liquid phase and 0.26% for the fluid phase.

4.11 Ternary System: CO_2 – Triglycerides – Free Fatty Acids

The ternary phase equilibrium for a CO_2 – TG – FFA system at 333 K and 20–30 MPa was predicted using the pure component parameters and the binary interaction parameters for the RKA model. The solute–solute interaction parameters between palm oil components (TG and FFA) were assumed to be zero. The reason behind this was that experimental data for the TG-FFA (for palm oil) was not available; hence, binary interaction parameters between these components could not be estimated. The phase equilibrium of a complex oil mixture is predicted based on only supercritical solvent–solute interactions while neglecting the solute–solute interactions.

Figure 4.22 provides a comparison between the ternary data for a CO_2-triolein-oleic acid system reported by Bharath et al. (1992) and the results obtained in this work. The phase behavior measured for this system showed that oleic acid and triolein are completely soluble, while the oleic acid-CO_2 and triolein-CO_2 pairs show only limited solubility. At all pressures measured, the solubility of oleic acid is higher than triolein. The average AAD for CO_2 composition in liquid phase and supercritical fluid phase is 4.06 and 0.25%, respectively. These values indicate that the RKA-predicted ternary phase diagrams are in good agreement with experimental data of Bharath et al. (1992). The assumption of negligible solute–solute interaction (in this case, interaction between fatty acid-triglyceride) is thus validated.

4.12 Distribution Coefficients of Palm Oil Components

The composition of palm oil components in the liquid and supercritical fluid phase can be described using the distribution coefficient, K_i:

$$K_i = \frac{x_i}{y_j} \tag{4.27}$$

where x_i, y_i is the mass fraction of component i in the liquid and supercritical fluid phase, respectively.

Figures 4.23 and 4.24 provide the distribution coefficients of palm oil components predicted using the RKA model, and those reported in the work of Gast et al. (2001), respectively. Note that the FFA and α-tocopherol are enriched in the fluid phase, whereas TG and β-carotene are enriched in the liquid phase. K_i was calculated on a CO_2-free basis since CO_2 is completely removed after the extraction process. An example calculating the distribution coefficients of palm oil components is given in Appendix B.

With respect to the distribution coefficient of α-tocopherol, discrepancy exists between the experimental value of Gast et al. (2001) and the value predicted based on the RKA-EOS. Contrary to our predicted value, Gast et al. (2001) reported that the fatty acids in a palm oil mixture are more soluble than tocopherol in supercritical CO_2. However, Škerget et al. (2002) and Stoldt and

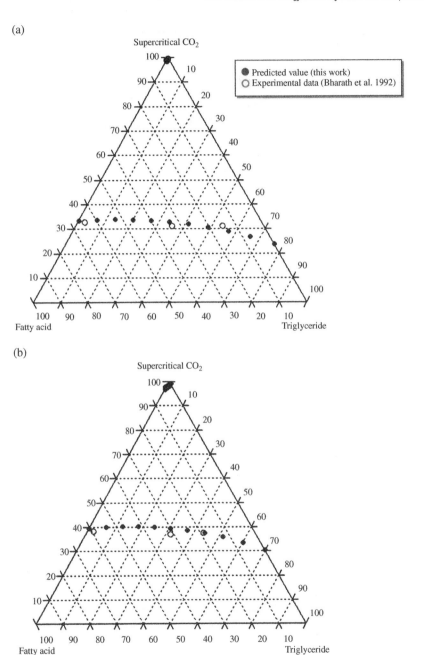

Figure 4.22 Phase equilibrium for a pseudo-ternary CO_2 – TG – FFA system at 333 K and (a) 20 MPa and (b) 30 MPa.

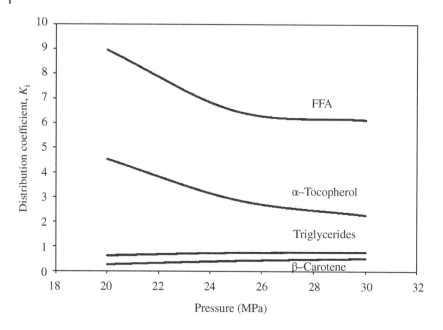

Figure 4.23 Predicted distribution coefficients (this work) for palm oil components at 343 K.

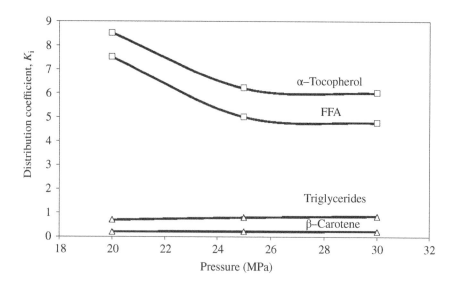

Figure 4.24 Experimental distribution coefficients (*Source:* Gast et al. (2001). © John Wiley & Sons) for palm oil components at 343 K.

Brunner (1998) observed that α-tocopherol has a smaller distribution coefficient than FFA in either milk thistle seed oil-supercritical CO_2 system or a palm deodorizer condensates-supercritical CO_2 system, respectively. These results are consistent with our findings.

The distribution coefficients for the FFA and TG provide information on the removal of FFA from the palm oil mixture. It can be observed from Figures 4.25 and 4.26 that triglycerides have a smaller distribution coefficient than FFA. The distribution coefficient of FFA varies in the range from 3 to 12, while the distribution coefficient of TG is in the range of 0.5–0.95. As pressure increases, the K-value of FFA decreases while the opposite trend is observed with regard to the K-value of palm oil TG. Note that a K-value of FFA that is greater than 1.0 means that FFA are enriched in the extract phase (supercritical CO_2-rich phase). A distribution coefficient for TG of less than 1.0 indicates that TG are enriched in the raffinate phase. A higher K-value for FFA compared to that for TG indicates that FFA has a higher solubility in supercritical CO_2.

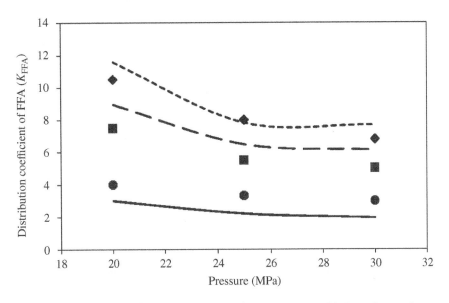

Figure 4.25 Distribution coefficients of FFA (CO_2-free basis). Experimental data (*Source:* Klein and Schulz (1989). © American Chemical Society): (♦) 370 K; (■) 340 K; (●) 310 K; Predicted results (this work): (---) 370 K; (--) 340 K; (- -) 310 K.

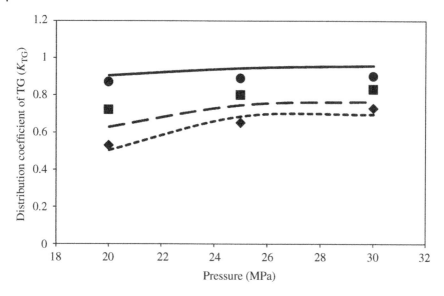

Figure 4.26 Distribution coefficients of TG (CO_2-free basis). Experimental data (*Source:* Klein and Schulz (1989). © American Chemical Society): (♦) 370 K; (■) 340 K; (●) 310 K; Predicted results (this work): (---) 370 K; (--) 340 K; (- -) 310 K.

4.13 Separation Factor Between Palm Oil Components

Separation factor is calculated from the respective compositions of fatty acids and triglycerides in the liquid and vapor phases. The separation factor, α_{ij}, is also defined as the ratio of distribution coefficients between two components (i, j):

$$\alpha_{ij} = \frac{y_i x_j}{x_i y_j} = \frac{K_i}{K_j} \qquad (4.28)$$

This parameter is useful to access the capability of the supercritical solvent to separate two compounds at a given extraction condition. The separation factor indicates the thermodynamic feasibility of the separation between two components. Separation factors are also the limiting value necessary for calculating the theoretical stages of the number of transfer units needed for a specific separation (Gast et al. 2001).

4.13.1 Separation Factor Between Fatty Acids and Triglycerides

The separation factor between FFA and TG indicates the thermodynamic feasibility of the separation of FFA from palm oil. The separation factors between FFA and palm oil TG in CPO are available in the literature. Figure 4.27 compares the separation factor obtained from the experimental data of Gast et al. (2001), and those calculated at various temperatures using Eq. 4.28. An example calculating the separation factor according to Eq. 4.28 is provided in Appendix B.

The results show that the separation factors between FFA and palm oil TG predicted by the RKA model are in good agreement with the experimental data. The average deviation was 6.62% over a wide range of temperatures (313–373 K) and pressures (20–30 MPa). This confirms that the model is a reliable alternative for predicting the distribution of palm oil components in supercritical CO_2 at various extraction conditions.

As can be seen, supercritical CO_2 is highly selective toward FFA compared to TG; with separation factors ranging from 2 to 20. Moreover, the separation

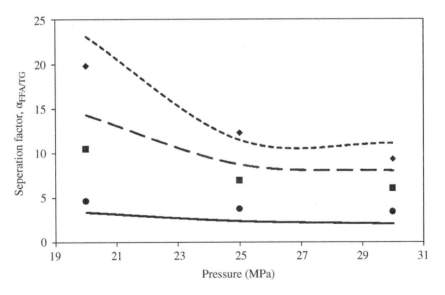

Figure 4.27 Separation factor between FFA and palm oil TG in a CPO–CO_2 system. Experimental data (*Source:* Gast et al. (2001). © John Wiley & Sons): (♦) 370 K; (■) 340 K; (●) 310 K. Predicted results (this work): (---) 370 K; (– –) 340 K; (– –) 310 K.

factor of FFA to TG increases with temperature and decreases with pressure. This trend has been observed in other similar systems. An increase in the extraction pressure causes a decrease in the enrichment of the more volatile substances in the vapor phase due to higher solvent capacity of carbon dioxide. The influence of the temperature can be explained by its effect on the vapor pressures of the mixture components. With increasing temperature, the relative change in the vapor pressures is higher for the fatty acids than for the triglycerides.

4.13.2 Separation Factor Between Fatty Acids and α-Tocopherols

The selectivity of supercritical CO_2 toward the fatty acids and α-tocopherols is also better seen by the respective separation factor. The experimental separation factors between FFA and α-tocopherols are available from the published data of Stoldt and Brunner (1999) for pseudo-binary system of supercritical CO_2 and complex mixtures of palm oil deodorizer condensates containing fatty acids and tocopherols. Figure 4.28 compares the separation factors obtained from the experimental data of Stoldt and Brunner (1999) and those calculated

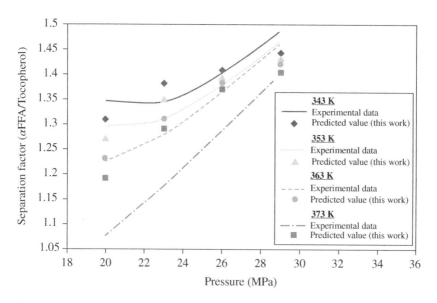

Figure 4.28 Comparison between the experiment data of Stoldt and Brunner (1999) and the calculated separation factor between FFA and α-tocopherols in mixtures of PFAD containing 6 wt% of tocopherols (*Source:* Stoldt and Brunner (1999). © Elsevier).

in this work at various operating conditions. The average deviation of the calculated separation factor from those reported in the literature (Stoldt and Brunner 1999) is 3.03%. In view of the small concentration of tocopherols, the predicted distribution coefficients are thus considered reliable.

4.14 Base Case Process Simulation

The base case study of the SFE process was performed to investigate the possibilities of using supercritical CO_2 to extract fatty acids from palm oil while recovering the valuable minor components present in palm oil. The simulation results were compared with the pilot plant data published in the literature to validate the developed process model. The base case process simulation results are described next.

4.14.1 Palm Oil Deacidification Process

To deacidify edible oils, supercritical CO_2 is able to separate TG and FFA since their vapor pressures and solubilities are different (Stahl et al. 1988). Supercritical CO_2 has been proven as a viable solvent to deacidify the CPO in order to produce refined quality palm oil (Brunner and Peter 1982; Ooi et al. 1996). In this study, the objective of deacidification column was to reduce the FFA content of the palm oil, thereby increasing the concentration of low volatile components in the raffinate phase product.

4.14.1.1 Solubility of Palm Oil in Supercritical CO_2

Figure 4.29 shows the predicted solubility of palm oil components in supercritical CO_2 (under equilibrium condition) through simulation at various operating conditions. It was observed that at all temperatures, the solubilities of palm oil were found to increase with pressure, which is consistent with the findings of Markom et al. (1999).

An increase in system pressure resulted in increase in solvent density which in turn increased solubility of a solute in the supercritical fluid. On the other hand, an increase in temperature lowered the solvent density, ultimately resulting in the opposite solubility behaviour. However, the effect of temperature on solubility was more complex since it did not only result in a decrease in solvent density but also an increase in the vapor pressure of the component, which tended to increase the solubility of a solute.

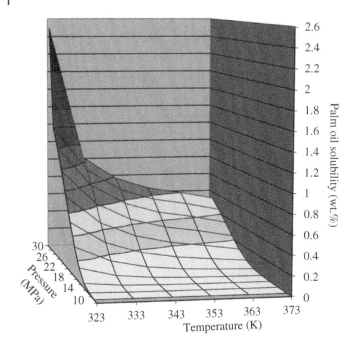

Figure 4.29 Effect of temperature and pressure on the solubility of palm oil in supercritical CO_2 (at S/F ratio of 40).

4.14.1.2 Palm Oil Deacidification Process: Comparison to Pilot Plant Results

Figure 4.30 compares the experimental solubility data of CPO in supercritical CO_2 determined experimentally by Ooi et al. (1996) to that predicted in this work. With reference to Figure 4.8, an extraction efficiency of 60% was assumed for the countercurrent extraction process with S/F ratio of 40. Less supercritical CO_2 flow rate was therefore required under continuous operation which uses 60% of the solvent that was required under equilibrium conditions. An example for calculating the palm oil solubility in supercritical CO_2 is given in Appendix C.

Ooi et al. (1996) adapted the simple countercurrent extraction scheme in an attempt to produce a refined quality palm oil. In their countercurrent scheme, CPO was continuously fed into the top of the extractor at a rate of 60 g/h, and supercritical CO_2 was pumped into the bottom of the column at a rate of 2400

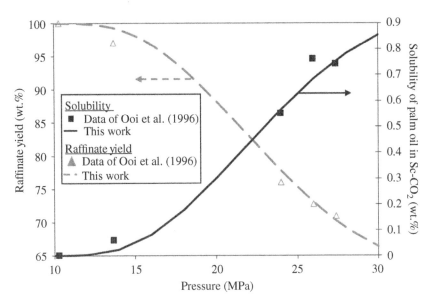

Figure 4.30 Effect of pressure on the yield of raffinate and solubility of palm oil in supercritical CO_2 at 50 °C (S/F ratio of 40 ± 5).

g/h. A simple extractor-separator system was rigorously simulated using the conditions from the pilot plant studies reported by Ooi et al. (1996).

Table 4.12 compares the simulation results and the experimental work of Ooi et al. (1996). The maximum deviation of the predicted FFA content in raffinate phase from the experimental data was 0.12 wt%. Larger deviation for the calculated yield was observed at high operating pressure of 27.4 MPa. This may be due to the low-density difference between coexisting liquid phase and fluid phase which makes the countercurrent operation infeasible at this operating condition (27.4 MPa, 323 K). Thus, the actual separation performance at low-density difference between supercritical CO_2 phase and liquid phase could not be well predicted by the model.

Gast et al. (2001) performed a study on the separation of palm oil components in a pilot-scale countercurrent packed column, which consists of both enriching and stripping sections. In the experiment carried out by Gast et al. (2001), CPO feed flux was kept between 71 and 82 g/h, and the mass flow rate of supercritical CO_2 was varied between 2.0 and 4.5 kg/h. This study implemented the same operating conditions used by Gast et al. (2001) to rigorously simulate the extractor. A comparison of the results obtained from

Table 4.12 Comparison of results obtained from simulation (this work) and experiment conducted by Ooi et al. (1996) for countercurrent extraction of palm oil at 323 K.

					FFA content of the raffinate			
Pressure (MPa)	27.4		24.0		13.7		10.3	
S/F ratio	62.8		58.2		88.7		40.0	
	Experimental data	Simulation	Experimental data	Simulation	Experimental data	Simulation	Experimental data	Simulation
FFA (%)	0.25	0.13	0.19 ± 0.01	0.23	1.21	1.43	2.02	2.33
Carotene (ppm)	—	865	690 ± 10	735	—	549	—	540
Yield (%)	72.0	50.45	68.5 ± 3.0	64.26	98.0	98.2	99.5	99.98
Extraction efficiency[a] (%)	—	57	—	58	—	53	—	63

[a] Refer to Figure 4.8.
Source: Modified from Ooi et al. (1996). © John Wiley & Sons.

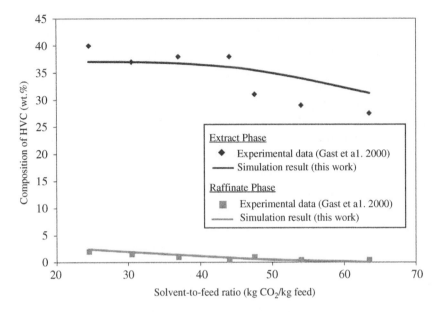

Figure 4.31 Countercurrent SFE of CPO (without reflux): experimental of Gast et al. (2001) and simulation results (this work) (*Source:* Gast et al. (2001). © John Wiley & Sons).

simulation of the countercurrent extraction process (without reflux) with that obtained from experiment conducted by Gast et al. (2001) at 370 K and 25 MPa, is shown in Figure 4.31. The content of high volatile components (i.e. FFA and tocochromanols) in the extract and raffinate phases is shown as a function of the solvent-to-feed ratio. Fatty acid rich fractions were recovered at the top of the column and deacidified palm oil was obtained at the bottom of the column. The average deviations of calculated FFA concentration in the extract and raffinate phases to the experimental data were 8.83 and 8.36%, respectively.

4.15 Conclusion

Due to the favorable properties of supercritical CO_2 mentioned earlier, this solvent has a great potential for CPO refining and for the recovery of valuable minor components. Modeling of high pressure phase equilibria in SFE processes is the key to reduce expensive and time-consuming high pressure experiments in the development of a new SFE process.

Pure component parameters for thermally labile palm oil components were estimated using the method proposed by Dohrn and Brunner (1994). Phase equilibrium calculations for palm oil components-supercritical CO_2 mixtures were made possible by introducing mixing rules for pure component parameters and by using the binary interaction parameters. Binary phase equilibrium data and solubility data for palm oil components with supercritical CO_2 in the literature were used for binary interaction parameters correlation. High pressure equilibrium data between the oil components were not available in the literature; therefore, only the interaction parameters between the palm oil components, and CO_2 were considered in the calculation procedure. The ultimate potential of the developed model needed to be determined by comparing experimental measurements to predictions for a wider range of SFE operating conditions.

The great variety of separation problems calls for suitable modeling and simulation tools for process synthesis and design. The development of a new flowsheet model for palm oil refining using SFE began with the synthesis of a conceptual flow sheet structure to recover high-purity palm oil and its minor components. Next, phase behavior of palm oil with supercritical CO_2 was observed using the RKA equation of state thermodynamic model available in the Aspen Plus® process simulation package. The application of temperature-dependent interaction parameters extended the predicting capability of the developed thermodynamic model. It is envisioned that the development of the new, intensified, and simpler palm oil refining process that is based on SFE technology can overcome the limitations of the existing technology for palm oil refining apart from making the refining process dramatically simpler along with the realization of the potentially profitable opportunities in the production of high added value products from oils and fats.

5

Application of Other Supercritical and Subcritical Modeling Techniques

5.1 Mass Transfer, Correlation, ANN, and Neuro Fuzzy Modeling of Sub- and Supercritical Fluid Extraction Processes

The mathematical modeling of experimental data has the objective to determine parameters for process design, such as equipment dimensions, solvent flow rate, and particle size, in order to make the estimation of the viability of SFE processes in industrial scale, through the simulation of the overall extraction curves (OECs). Because of the enormous variety of solid substratum that can be suitable for SFE, the modeling of the OECs can be done applying several models (Esquível et al. 1999; Marcelo et al. 2014). In this case, each author gives his/her own interpretation of the mass transport phenomena that happens during the process. The mass balance equations can be solved analytically depending on the assumptions made (Melreles et al. 2009) or they can be solved numerically. In any case, when the model describes the phenomena occurring in the extractor vessels, it can safely be used for process design and process scale-up. In order to solve the differential mass balance equations and to analyze the adequacy of the model, it is necessary to compare calculated OECs with experimental data.

Mass balance equations sometimes need to be followed by energy equations when the SFE system is not isothermal and temperature change along the length of the reactor or in radial coordinate is noticeable. In this case, mass and energy balance equations need to be solved simultaneously. Also, in cases which chemical reactions are involved in extraction heat transfer may be considered. The momentum balance equation in studying SFE can be considered when bed diameter is large compared to bed length which causes velocity profile. CFD study of the bed can yield to consideration of momentum balance

---

Modeling, Simulation, and Optimization of Supercritical and Subcritical Fluid Extraction Processes, First Edition. Zainuddin A. Manan, Gholamreza Zahedi, and Ana Najwa Mustapa.
© 2022 by the American Institute of Chemical Engineers, Inc.
Published 2022 by John Wiley & Sons, Inc.

equations. In this case, fluid mixing and velocity profiles along the extraction bed will be important for researchers (Fernandes et al. 2009).

For most of modeling applications, experimental design or using mass balance equations have been proven to be sufficient. Physics of the system, our understanding of the subcritical or supercritical system is crucial for model development. The general rule which can be suggested is that the researcher can start from simple mass balance equations. Comparison of the model with experimental data is the key to deciding if the proposed model is suitable or not. If the accuracy of the proposed model is not enough, then the model complexity should be increased. For example, if the bed height is 2 m and bed diameter is 0.5 m, first, one-dimensional mass balance with change of mass along the bed height can be employed. If the model accuracy is not acceptable, then radial variation should be included and the concentration change in both axial and radial coordinates should be found. If the results are not satisfactory, the researcher needs to check the assumptions, which have been made for model development. For example, velocity may not be constant and the velocity change with axial and radial coordinates should be considered. This method is a general technique which starts with simplifying and relaxing assumptions. Based on model accuracy the model becomes more complex in next steps until the satisfactory accuracy of the model is achieved.

This chapter presents the applications of a variety of modeling approaches for sub- and supercritical fluid extraction (SFE) processing of essential and edible oils. Sections 5.2–5.4 of this chapter describe the applications of mass transfer model, correlation model, ANN model, and neuro fuzzy model. In Sections 5.5 and 5.7, the ANFIS and Grey Box modeling approaches are applied for SFE of Anise Seed. Sections 5.8–5.11 demonstrate and compare the pros and cons of statistical and ANN SFE modeling for a variety of oils.

5.2 Mass Transfer Model

This model is very common in most sub- and supercritical modeling. Case study is the SFE from the root of Vetiver because of its commercial importance. Vetiver (Vetiveria zizanioides (L.) Nash ex Small) is found in tropical regions of the planet, such as India, China, Indonesia, Haiti, and the Reunion Island, which are the world main vetiver oil producers. Brazil and other South American countries have an incipient production of vetiver. The volatile oil from vetiver roots is a viscous liquid at ambient temperature. Its color varies from amber to dark brown, and whose odor has sweet, earthy, and woody notes (Melreles et al. 2009).

5.2 Mass Transfer Model

Figure 5.1 depicts schematic of SFE bed. The fixed bed consists of vetiver roots as the stationary phase with the flowing supercritical carbon dioxide as the mobile phase. Solute (SFE extract) transfers from solid to the bulk of supercritical fluid. The quality of the model depends on the assumptions considered. The assumptions are the following:

1) Radial dispersion is neglected (one-dimensional approach).
2) Radial concentration gradients are neglected.
3) The system is isothermal and isobaric (so heat transfer equations are not needed).
4) The physical properties of the supercritical fluid are constant.
5) There is local equilibrium at interface of fluid and solid phase.
6) Velocity is constant along the bed (so momentum balance equations are not needed).

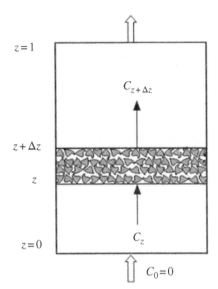

Figure 5.1 Schematic of element in SFE (*Source:* Melreles et al. (2009). © Elsevier).

These assumptions will yield only in mass balance equations in cylindrical coordinate. Applying conservation of mass, the following equation can be written for element in Figure 5.1:

Mass balance for fluid phase:

$$\varepsilon\left(|C.V.A|_z - |C.V.A|_{z+\Delta z}\right) + \varepsilon\left(|j_A.A|_z - |j_A.A|_{z+dz}\right)$$
$$+ [A.k_f.(C_s - C).(1-\varepsilon)] = \frac{C|z+\Delta z - C|z}{\Delta t}.\varepsilon.\Delta z.A \quad (5.1)$$

which will yield to Eq. 5.2 as the following:

$$\frac{\partial C}{\partial t} + V\frac{\partial C}{\partial z} = D_1 \frac{\partial^2 C}{\partial z^2} + \frac{3k_f}{R_p}\left(\frac{1-\varepsilon}{\varepsilon}\right)(C_s - C) \quad (5.2)$$

By defining $Pe_b = \frac{L.V}{D_1}$ and dimensionless length $Z = \frac{z}{L}$ and time $\tau = \frac{t.V}{L}$:

$$\frac{\partial C}{\partial \tau} - \frac{1}{Pe_b}\frac{\partial^2 C}{\partial z^2} + \frac{\partial C}{\partial z} + \frac{1-\varepsilon}{\varepsilon}\frac{3L}{R_p}\frac{Bi}{Pe_p}(C - c_s) = 0 \quad (5.3)$$

The boundary conditions for Eq. 5.3 are the following:

at $z = 0 \Longrightarrow C = 0$

at $z = 1 \Longrightarrow \dfrac{\partial C}{\partial z} = 0$

at $\tau = 0 \Longrightarrow C = 0$

Applying mass balance for the solid phase will result in Eq. 5.4:

$$\frac{\partial c}{\partial \tau} = \frac{1}{Pe_p} \frac{L}{R_p} \frac{1}{\rho^2} \frac{\partial}{\partial \rho}\left[\rho^2 \frac{\partial c}{\partial \rho}\right] \tag{5.4}$$

ρ is dimensionless variable of r/R_p. This equation is mass balance inside spherical coordinates considering only radial coordinates. The relevant boundary conditions for Eq. 5.4 are the following:

at $\rho = 1 \Longrightarrow Bi(C - c_s/k) = \dfrac{\partial c}{\partial \rho}$

at $\rho = 0 \Longrightarrow \dfrac{\partial c}{\partial \rho} = 0$

at $\tau = 0 \Longrightarrow c = c_0$

where k is the equilibrium constant (ratio of solute concentration in the vetiver root particle to solute concentration in the fluid at equilibrium condition). All of these equations now should be solved to find concentration profile or extraction yield. If the results are accurate, then the model will be acceptable otherwise, the assumptions should be reduced and two-dimensional models should be considered. One of the important factors in solving extraction models is finding the value of equilibrium constant. The equilibrium constant can be found from previously developed correlations, thermo dynamical published data. Also, in our work, we have described a simple optimization method to find k value. Our developed correlation can be employed in SFE modeling as the following (Zahedi et al. 2010b):

$$Sh = 0.0306 \, Re^{0.75} \, Sc^{0.33} \tag{5.5}$$

To solve two partial differential equations, different numerical methods or different softwares can be employed. In the following, a finite difference method which can be implemented in different programming environments is described (Constantinides and Mostoufi 1999). In a finite difference method, the length of extractors divide into "h" segments and the radius of the particles also is divided into "m" parts, then the PDEs are expanded by finite difference methods. Figure 5.2 shows Fuzzy membership functions which will help facilitate SFE modeling in section 5.4. The following formulas are used in discretizing the differential equations:

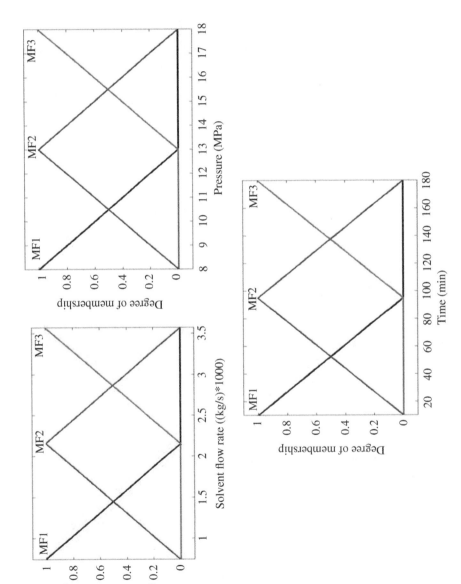

Figure 5.2 Membership functions for process variables.

$$\frac{dy(i)}{dZ} = \frac{y(i) + y(i-1)}{\Delta Z} \tag{5.6}$$

$$\frac{d^2y(i)}{dz^2} = \frac{y(i+1) - 2y(i) + y(i-1)}{\Delta Z^2} \tag{5.7}$$

For Eq. 5.3, the discretizing in z direction will yield to:

$$\frac{\partial y(i)}{\partial \tau} - \frac{1}{Pe_b} \frac{y(i+1) - 2y(i) + y(i-1)}{(\Delta z)^2} + \frac{y(i) - y(i-1)}{\Delta z}$$

$$+ \frac{1-\varepsilon}{\varepsilon} \frac{3L}{R_p} \frac{Bi}{Pe_p} (y(i) - y(h+im)/k) = 0 \tag{5.8}$$

where $i = 2, 3, 4, \ldots, h-1$ and $j = 2, 3, \ldots, m-1$

The boundary conditions for this equation are

at $z = 0 \Longrightarrow C = 0 \Longrightarrow y(1) = 0$

at $z = 1 \Longrightarrow \dfrac{\partial C}{\partial z} = 0 \Longrightarrow \dfrac{y(h) - y(h-1)}{\Delta z} = 0 \Longrightarrow y(h) = y(h-1)$

at $\tau = 0 \Rightarrow C = 0$

For Eq. 5.8,

$$\frac{\partial y(h + (i-1)m + j)}{\partial \tau} =$$

$$\frac{1}{Pe_p} \frac{L}{R_p} \left[\frac{y(h + (i-1)m + j + 1) - 2y(h + (i-1)m + j) + y(h + (i-1)m + j - 1)}{(\Delta \rho)^2} \right.$$

$$\left. + \frac{2}{(j-1)\Delta \rho} \frac{y(h + (i-1)m + j) - y(h + (i-1)m + j - 1)}{\Delta \rho} \right]$$

The boundary conditions for this equation are

at $\rho = 1 \Longrightarrow Bi\left(C - \dfrac{C_s}{k}\right) = \dfrac{\partial c}{\partial \rho} \Longrightarrow Bi(y(i) - y(h+im)/k)$

$$= \frac{y(h+im) - y(h+im-1)}{\Delta \rho}$$

$$\Longrightarrow y(h+im) = \frac{1}{\frac{1}{\Delta \rho} + \frac{Bi}{k}} \left[Biy(i) + \frac{1}{\Delta \rho} y(h+im-1) \right]$$

at $\rho = 0 \Longrightarrow \dfrac{\partial c}{\partial \rho} = 0 \Longrightarrow \dfrac{y(h + (i-1)m + 2) - y(h + (i-1)m + 1)}{\Delta \rho} = 0$

$$\Rightarrow y(h + (i-1)m + 1) = y(h + (i-1)m + 2)$$

at $\tau = 0 \Rightarrow c = c_0$

Now as it is obvious, the set of algebraic equations has appeared which solving them will result in a solution for the PDE model. The current approach showed acceptable results, and we did not need to make the model complex (Melreles et al. 2009). The proposed model showed satisfactory results in most of our SFE modeling and also other researcher modeling (Cavalcanti et al. 2016; Melreles et al. 2009; Rahimi et al. 2011; Zahedi et al. 2010b).

5.3 ANN Modeling

In this part, a case study of us has been demonstrated for better understanding of ANN modeling. The work deals with SFE of oil from Passiflora seed (Zahedi and Azarpour 2011). First, after data collection to make sure that the selected data for modeling present normal operating ranges, the unsatisfactory data were excluded from the data bank. The off data which were not in the normal operation range of the process were removed.

The back-propagation learning with one hidden layer network has been used in this work. Inputs and outputs are normalized between the values −1 and 1. Logistic Sigmoid and purelin transfer functions have been used in constructing ANNs. ANN has been trained with 70% of the data set, and 30% of the data have been applied for testing the predictions of ANN. The Levenberg–Marquart training algorithm was employed for modeling the oil extraction yield of Passiflora seed. There is not any general and precise method to achieve the optimum number of hidden layers of the neurons, and it is obtained by trial and error. The optimum number of hidden layer neurons was determined to be 11 for this network. Mean Square Error (MSE) values for ANN model is calculated 0.0009. MSE is defined as follows:

$$\text{MSE} = \frac{\Sigma(X_{\exp} - X_{\text{sim}})}{n} \tag{5.9}$$

5.4 Neuro Fuzzy Modeling

In this part, an example of using neuro fuzzy in SFE is presented (Davoody et al. 2012). In this work, after getting data for extracting Anise from Anise seeds, ANFIS network with Sugeno-style inference system was designed which maps three independent variables as input data to mass of extract as output. The input variables are time, temperature, and pressure. Three Triangular membership functions have been considered for each input data. Figure 5.2 shows the final and best-obtained membership functions.

Comparison of the developed neuro fuzzy model with unseen data shows that the obtained model reaches the test relative error of 1.98 which confirms the excellent capability of neuro fuzzy modeling. The following equation corresponds to mean particle diameter:

$$\bar{d} = \frac{1}{\sum_{i=1}^{k} \frac{\Delta X_i}{d_i}} \quad i = 1, 2, 3, ..., k, \tag{5.10}$$

where the diameter of the sieve is shown by d_i, and X_i is the mass retained by the sieve. This diameter can be used in modeling where different seeds with different diameters exist. The following equation is used to calculate porosity (g) of the bed beside the particles:

$$\varepsilon = 1 - \frac{\rho_a}{\rho_r} \tag{5.11}$$

5.5 ANFIS and Gray-box Modeling of Anise Seeds

Pimpinella anisum L. Umbelliferae is the botanical name of Anise which belongs to the Apiaceae (also known as Umbelliferae) plant family. It is an annual herb indigenous to Iran, India, Turkey, and many warm regions in the world (Zargari 1989). This plant is a dainty herbaceous plant with white flowers. The main components of Anise are the following: polyacetylenes, anethole, eugenol, umbelliferon, pseudoisoeugenol, estragole, methylchavicol, anisaldehyde, terpenehydrocarbons, estrols, scopletin, coumatins, and polynes (Gülcin et al. 2003). Anise fruit is known as aniseed; contains 1–4% of essential oil. Perfumery, medicine, and flavoring industries use essential oil of *P. anisum* L. as feedstock. Medicinal applications include its use as an appetizer, carminative, and sedative agent, or for stimulating milk production in breast-feeding mothers (Özel 2009). Pourgholami et al. in 1999 investigated the anticonvulsant effects of *P. anisum* essential oil. In 2001, Boskabady and Ramazani–Assari reported the bronchodilatory effects of the essential oil, the aqueous extract, and the ethanolic extract of Anise. The authors concluded that the relaxant effect of the plant is not due to an inhibitory effect of histamine, but instead due to inhibitory effects on muscarinic receptors.

In this study, first, Superciritcal Fluid Extraction (SFE) of *P. anisum* L. seed is modeled using ANFIS technique. Thereafter, the development of an artificial intelligence modeling scheme using the ANFIS methodology was described. The proposed neuro-fuzzy model deals with pressure, solvent mass flow rate, and time data as the input variables, which are readily available for most of the extractions.

In the next part of the study, a gray box model was proposed. In this case, a white box model is designed. Then a neuro-fuzzy network to estimate Sherwood number which served as a black box was developed and was linked to the white box model. This hybrid or gray box model was utilized for performance analysis of the SFE process.

5.6 White Box SFE Modeling of Anise

In this part, our previously developed model is employed as white box model (Melreles et al. 2009). Partial differential equations (PDEs) are used to describe the extraction rate. Eight assumptions are made to reach the material balance equation (Shokri et al. 2011). Based on the hypothesis of the model, the following material balance equation is obtained (Zahedi et al. 2010a).

$$\frac{\partial C_i}{\partial \tau} - \frac{1}{Pe_b}\frac{\partial^2 C_i}{\partial z^2} + \frac{\partial C_i}{\partial z} + \frac{1-\varepsilon}{\varepsilon}3\frac{k_f}{v}\frac{L}{R_p}(C_i - C_{i,s}) = 0 \qquad (5.12)$$

Initial and boundary conditions are as follows:

$$\text{at } z = 0^- \quad C_i = 0 \qquad (5.13a)$$

$$\text{at } z = 0^+ \quad \frac{\partial C_i}{\partial z} = Pe_b(C_i - 0) \qquad (5.13b)$$

$$\text{at } z = 1 \quad \frac{\partial C_i}{\partial z} = 0 \qquad (5.14)$$

$$\text{at } \tau = 0 \quad c_i = 0 \qquad (5.15)$$

where 0^+ and 0^- refer to the place before and after of the extractor entrance, respectively. The following equation is generated for the particles, which is followed by its initial conditions:

$$\frac{dq_i}{d\tau} = -3\frac{k_f}{v}\frac{L}{R_p}(C_{i,s} - C_i) \qquad (5.16)$$

$$\text{at } \tau = 0 \quad q_i = q_{i,0} \qquad (5.17)$$

C_s and q are related by the following equilibrium:

$$q_i = KC_{i,s} \tag{5.18}$$

Equations 5.2–5.18 were solved numerically in MATLAB software using numerical methods reported in our previous work (Melreles et al. 2009).

5.6.1 Gray Box Parameters

Density and viscosity of pure CO_2 were obtained from the NIST Chemistry WebBook (NIST). Cathpole and King correlation was employed to obtain the binary diffusion coefficient of solute in the supercritical solvent (Catchpole and King 1994). The axial dispersion coefficient was obtained using Funazukuri et al. correlation as the following (Funazukuri et al. 1998):

$$\frac{\epsilon D_i}{D_m} = 1.317(\epsilon \text{ Re Sc})^{1.392} \tag{5.19}$$

The external mass transfer coefficient was calculated by the following equation:

$$\text{Sh} = \frac{2R_p k_f}{D_m} \tag{5.20}$$

In order to calculate mass transfer coefficient, Sherwood number has to be determined. The method of determining Sherwood number is the key difference between mathematical models and gray box. In order to estimate Sherwood numbers in gray box, a neuro-fuzzy network is trained based on experimental values of Reynolds and Schmidt numbers. Once the Sherwood number is estimated by the designed model, mass transfer coefficient can be easily calculated by Eq. 5.20. The values of Sherwood number would be more accurate if they are estimated by a reliable model instead of being calculated by a mathematical correlation. This designed model serves as the black box, and as it gets combined by white box (mathematical model), gray box is designed. The rest of the parameters are calculated in the white box, and the final values of all parameters (including the one estimated by black box) enter the gray box for the final result.

5.6.2 ANFIS

Operational conditions for all four experiments are reported in Table 5.1. In Table 5.1, D_p, C, and x_0 refer to particle diameter (m), oil concentration in supercritical fluid (kg/m^3), and initial mass fraction of extractable oil in solid phase, respectively.

Table 5.1 Operational condition for four experiments.

Conditions	Exp. 1	Exp. 2	Exp. 3	Exp. 4
T(K)	303.15	303.15	303.15	303.15
P(MPa)	8	10	14	18
Q (kg/s) $\times 10^5$	1.23–5.17	1.35–5.95	1.82–4.45	2.33–4.6
D_p(m) $\times 10^4$	7.95	7.95	7.95	7.95
C (kg/m^3)	1321	1321	1321	1321
x_0(g extract. ganise^{-1}) $\times 10^2$	3.13	7.94	10.48	10.67
ε(-)	0.63	0.63	0.63	0.63
L(m)	0.12	0.12	0.12	0.12
Mass of seed (kg)	0.191	0.191	0.191	0.191

SFE process takes place at a constant temperature and four different pressures. Solvent mass flow rates change in each pressure, and every 10 minutes, the accumulated mass of extract in each solvent flow rate is recorded. Since other parameters do not change during the experiment, ANFIS model has three inputs (pressure, solvent mass flow rate, and extraction time), and one output (mass of extract). The 369 observations from our experiments were used in this part. The 277 observations were used to train the network, and the remaining 92 observations were kept for further testing the developed network. Collection of data for training is done randomly and based on principle component analysis in order to make sure the collected data are true representative of all data.

5.6.2.1 Preprocessing

Preprocessing the input and output data can improve the performance of the model. The input data go under a specific process before entering the network. This process is meant to scale the data in a way that makes them more suitable and understandable for the network. In this study, principal component analysis is employed for preprocessing. This analysis is a procedure for scaling of the input vectors which are very effective when components of the input vectors are highly correlated (redundant) (Treybal 1990).

5.6.3 Gray Box

Shokri et al. have reported the values of dimensionless parameters during modeling (Shokri et al. 2011). The experimental values of Re, Sc, and Sh numbers mentioned in that manuscript are employed in this study to design a neuro-fuzzy network. The data set includes 20 sets of data. A total of 75% of the whole data (15 sets) are used for training networks and remaining 25% (5 sets) are used for testing. The designed model is meant to estimate Sh numbers based on Re and Sc numbers. Table 5.2 shows the available data set.

Table 5.2 Available data of Re, Sc, and Sh numbers.

Re	Sc	Sh × 10^3
0.09	4.77	8.14
0.18	4.77	14.19
0.23	4.77	17.27
0.24	4.77	17.54
0.36	4.77	23.91
0.08	6.02	8.33
0.11	6.02	10.68
0.14	6.02	12.75
0.21	6.02	16.91
0.21	6.02	25.35
0.09	7.56	9.99
0.12	7.56	12.10
0.17	7.56	15.5
0.18	7.56	16.55
0.23	7.56	19.53
0.11	8.77	11.7
0.15	8.77	15.27
0.16	8.77	15.71
0.21	8.77	19.49

5.7 Results and Discussion

5.7.1 ANFIS

As discussed before, three parameters were selected for input variables (i.e. solvent mass flow rate, time, and pressure). Based on these variables, an ANFIS network with Sugeno-style inference system has been designed which maps three independent variables as input data to mass of extract as output. Three Triangular membership functions have been considered for each input data. Figure 5.3 shows the final and best obtained membership functions.

The values of the designed model's outputs were compared to those of experimental data in Figure 5.4. The solid lines represent the model's predictions, while blue markers stand for experimental data. These figures confirm good overlap between model outputs and experimental data.

Figure 5.4 compares experimental data with ANFIS outputs in pressure of 18 MPa. Linear data in the flow rate of 2.33×10^{-5} kg/s, result in accuracy of ANFIS predictions.

5.7.2 Gray Box Modeling Results

5.7.2.1 Black Box

Two statistical indices (RMSE and AAD%) are employed to verify the performance of the network. Table 5.3 presents the statistical performance of the designed ANFIS model for gray boxes.

The errors mentioned in Table 5.3 demonstrate ANFIS network has excellent performance in catching relationships between input and output data although a few sets of reliable data were available. It was verified that "Gaussian curve built-in" was the best type of membership function for both inputs.

Model estimations were compared with those of real experiments in four experiments. Operational conditions of the experiments have been tabulated in Table 5.1. Figures illustrate the comparison.

In Figure 5.5, Gray box results are in good agreement with experimental data in the first flow rate. Like ANFIS, gray box model shows inaccuracy in the last flow rate where data are highly nonlinear. Gray box results are much closer to the targets compared to mathematical model results. Acceptable agreement was observed between model predictions and experimental data. Obviously, more accurate Sh numbers have a positive impact on the performance of the new model.

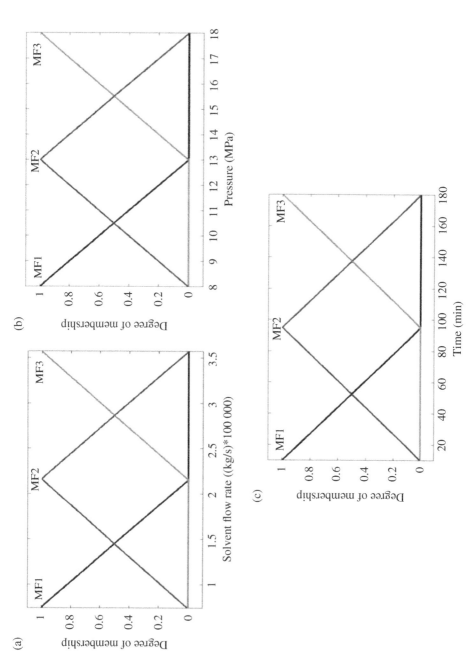

Figure 5.3 Membership functions of (a) Solvent flow rate, (b) Pressure, and (c) Time.

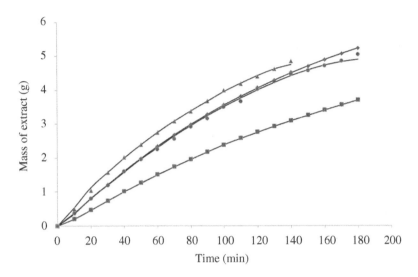

Figure 5.4 Extraction from Anise-comparison between NF model results (--) and the experimental data obtained in Exp. 4 at solvent mass flow rate of 2.33 × 10^{-5} kg/s (■), 3.32 × 10^{-5} kg/s (●), 3.45 × 10^{-5} kg/s (◆), and 4.60 × 10^{-5} kg/s(▲) (*Source: Davoody (2012). © John Wiley & Sons*).

Table 5.3 Statistical performance of a designed black box.

During training				During test			
RMSE	MSE	R	AAD%	RMSE	MSE	R	AAD%
0.000 099 8	9.9 × 10^{-9}	0.999 79	0.5857	0.000 48	2.3 × 10^{-7}	0.998 93	2.4973

5.7.3 Comparison of ANFIS and Gray Box Models with ANN and White Box Models

Shokri et al. have applied mathematical and ANN modeling on this (Shokri et al. 2011). In this part, results of all four models (including the two proposed models in this study) are compared. Table 5.4 presents RMSE values of each model in different solvent flow rates in each experiment. In Table 5.4, RMSE1,

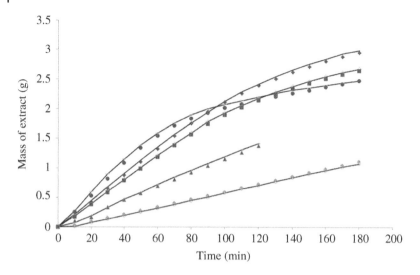

Figure 5.5 Extraction from Anise-comparison between GB model results (--) and the experimental data obtained in Exp. 1 at solvent mass flow rate of 1.23×10^{-5} kg/s (◇), 2.58×10^{-5} kg/s (▲), 3.35×10^{-5} kg/s (◆), 3.42×10^{-5} kg/s (■), and 5.17×10^{-5} kg/s (●) (*Source*: Davoody (2012). © John Wiley & Sons).

2, 3, and 4 refer to RMSE of mathematical model, ANN, Gray Box, and ANFIS model results, respectively.

Table 5.4 depicts estimation of ANFIS and Gray box models are more accurate than those of ANN and mathematical models. Obviously, equipping a mathematical model with neuro-fuzzy network has improved performance of the model significantly. Even in some cases, gray box outputs are more accurate than ANN outputs.

5.8 Introduction – Statistical versus ANN Modeling

During the course of conducting SFE research, it is important to know the limitations and capabilities of the tools for data mapping. Traditional Design of Experiment (DOE) tools are very well known, and they are easy to use. There is user friendly software, like JMP, which enables users to develop DOE and statistical modeling with limited statistical knowledge. These models need less data compared to Artificial Intelligence (AI) models. Majority of AI based

Table 5.4 Statistical performance of all four models.

Experiment	P (MPa)	Q (kg/s × 10⁻⁵)	RMSE 1	RMSE 2	RMSE 3	RMSE 4
1	8	1.23	0.0321	0.0279	0.0213	0.0024
	8	2.58	0.0357	0.0581	0.0298	0.0426
	8	3.35	0.0578	0.0414	0.0313	0.0028
	8	3.42	0.0666	0.0298	0.0313	0.0026
	8	5.17	0.0703	0.0502	0.0541	0.0493
2	10	1.35	0.0221	0.0305	0.0219	0.0048
	10	1.88	0.0716	0.0243	0.0255	0.0159
	10	2.38	0.0998	0.0404	0.0250	0.0182
	10	3.47	0.1266	0.0573	0.0445	0.0286
	10	3.55	0.1188	0.0760	0.0661	0.0424
	10	5.95	0.0958	0.1283	0.0852	0.0320
3	14	1.82	0.0670	0.0350	0.0192	0.0031
	14	2.35	0.0926	0.0355	0.0286	0.0158
	14	3.27	0.0753	0.0276	0.0331	0.0106
	14	3.57	0.0642	0.0413	0.0382	0.0065
	14	4.45	0.1033	0.0137	0.0212	0.0153
4	18	2.33	0.0649	0.0206	0.0200	0.0045
	18	3.32	0.1141	0.0480	0.0566	0.0444
	18	3.45	0.0773	0.0323	0.0360	0.0049
	18	4.6	0.0841	0.03	0.0438	0.0398

models are poor in extrapolation, while statistical models are fine for extrapolations. Statistical models provide equation/s which researchers can share, while developed AI models are like a box which only can be shared with different users. Training process for AI models is time-consuming.

AI models are very accurate. If the number of output (response) increases, AI models are the best choices for data modeling. AI models are more robust to noisy data and regardless of faulty and noisy data, they can provide accurate outputs. Definity optimization and control of the processes with very accurate models will yield better results so AI models have superiority in predictions accuracy.

In this chapter, three case studies will be presented. The first case is supercritical carbon dioxide extraction of *Quercus infectoria* (manjakani) oil by response surface methodology. This is a very interesting study on the comparison between ANN and DOE technique. The second study is on the subcritical fluid extraction of *Orthosiphon stamineus* oil which is a local herb in Malaysia. The final study is on the SFE of essential oil from passion fruit seed. The study provides a good comparison between statistical modeling and ANN modeling.

5.9 Supercritical Carbon Dioxide Extraction of *Q. infectoria* Oil

Q. infectoria is a small tree native of Greece, Turkey, and Iran with 4–6 feet in height. In Malaysia, it is a small tree, shrubby looking with smooth and bright-green leaves borne on short petioles of 1–1.5 inches long. Manjakani is used by countries extended from Greece, Turkey, Syria, to China and Malaysia for medicinal purpose. The best-quality *Q. infectoria* can be found in Aleppo, Syria.

Q. infectoria or *Aleppo tanning* is used in treating vaginal discharge, curing sore throat, ulcer, and skin disease. *Q. infectoria* contains *astringent* which has anti-bacteria properties. Manjakani also can act as an anti-microbe and anti-inflammatory. In addition, *A. tanning* could be used for curing *chronic dysentery, diarrhea, passive hemorrhages*, and in cases of *poisoning by strychnine* because of its anti-bacteria properties (Baharuddin et al. 2015).

As discussed earlier, there are several approaches to optimize the extraction process. Pressure, temperature, extraction time, sample size, and cycle time are among the operating parameters which can optimize an extraction process. Those strategies are classified as statistical optimization. One approach for statistical optimization is design of experiment, which numerous factors are varied during the designing process. In this study, the experimental RSM with full factorial composite design involving three factors namely: temperature, pressure, and extraction time has been considered. In addition, three levels were applied in order to obtain a second-order polynomial model for the lycopene maximization.

5.9.1 Materials and Methods

Aleppo tanning galls were obtained from the local market in Johor, Malaysia. They were dried in an oven at 45 °C for 24 hours and then were ground into powder to 200 μm in size. As the size is fixed, so the effect of particle diameter will not be studied at this research. The powder was sealed and stored in containers at room temperature. Pure carbon dioxide was used for supercritical extraction.

To start supercritical extraction 2 g of powered gall was used. A high-pressure pump was employed to pump liquid carbon dioxide. Temperature, pressure, and extraction time were among the parameters that we were able to change during the SFE process. After setting and recording desired operating parameters, the extracted oil was collected in vials. The oil was measured, and extraction time, corresponding temperature, and pressure were recorded. Equation 5.21 depicts how extraction yield was calculated after each trial:

$$\text{Oil yield} = (\text{mass of extracted oil}/\text{mass of dried material}) \quad (5.21)$$

5.9.2 Experimental Design

RSM was employed to correlate pressure, temperature, and extraction time to extraction yield. As the size of material was fixed, so it will not play a role in oil extraction. Pressure (X_1), temperature (X_2), and extraction time (X_3) were scales between −1 and 1 to start RSM analysis. Table 5.5 shows coded values and the range.

Equation 5.22 shows the second-order regression which has been used for data mapping at this study:

$$Y = \beta_0 + \Sigma \beta_i X_i + \Sigma \beta_{ii} X_i^2 + \Sigma\Sigma \beta_{ij} X_i X_j \quad (5.22)$$

Table 5.5 Levels of independent variables used in RSM design.

Symbol		Coded levels		
Factor levels	Independent variables	−1	0	1
X_1	Pressure (Psi)	5000	6000	7000
X_2	Temperature (°C)	40	50	60
X_3	Time (min)	30	45	60

Y is extraction yield, and β_{ij} are the interactive coefficients. For more details on statistical modeling, please refer to Chapter 4.

In order to start DOE, 17 run Box–Behnken Design including three levels and five replicates at the center point was implemented. Table 5.6 has summarized the design values.

According to Table 5.6, the maximum extraction yield (100% in theory) was achieved after 60 minutes extraction time and temperature of 60 °C and pressure of 6000 psia.

Table 5.6 Box–Behnken experimental design and validation.

Run	P (Psi)	T (°C)	t (Min)	Yield (%) (observed)	Yield (%) (predicted)	Standard deviation
1	5000	40	45	0.56	0.63	0.05
2	7000	40	45	0.69	0.65	0.03
3	5000	60	45	0.93	0.89	0.03
4	7000	60	45	0.89	0.90	0.01
5	5000	50	30	0.61	0.62	0.01
6	7000	50	30	0.64	0.63	0.01
7	5000	50	60	0.80	0.76	0.03
8	7000	50	60	0.74	0.77	0.02
9	6000	40	30	0.63	0.60	0.02
10	6000	60	30	0.84	0.86	0.01
11	6000	40	60	0.73	0.74	0.01
12	6000	60	60	1.00	1.00	0.00
13	6000	50	45	0.93	0.93	0.00
14	6000	50	45	0.93	0.93	0.00
15	6000	50	45	0.93	0.93	0.00
16	6000	50	45	0.94	0.93	0.01
17	6000	50	45	0.93	0.93	0.00

Table 5.7 shows second-order polynomial model parameter. Significance of each parameter has been checked using t-test. P-values have been depicted in this table.

The coefficients with ($p > 0.05$) are not dominant coefficients and were ignored in main equation form. After applying these corrections, the main equation to predict oil extraction as a function of extraction time, pressure, and temperature will have the following form:

$$Y = -7.9205 + 0.0019X_1 + 0.0651X_2 + 0.0490X_3 - 0.0003X_2^2$$
$$- 0.0005X_3^2 + 0.0001X_2X_3 \tag{5.23}$$

Analysis of variation (ANOVA) was employed to evaluate models to predict the SFE system result. As it is obvious from Table 5.8, the system has a good prediction performance with R^2 value equal to 0.991. The model predicts max

Table 5.7 Statistical model parameters and validation.

Model parameters	Regression coefficients	Standard error	t	Significance level (P value)
Constant				
X_0	−7.9205	0.5847	−13.546	0.000
Linear coefficient				
X_1, pressure	0.0019	0.0001	13.936	0.000
X_2, temperature	0.0651	0.0123	5.304	0.001
X_3, extraction time	0.0490	0.0068	7.250	0.000
Quadratic coefficient				
X_1^2	−0.0000	0.0000	−13.221	0.000
X_2^2	−0.0003	0.0001	−3.070	0.019
X_3^2	−0.0005	0.0000	−10.003	0.000
Interaction coefficient				
$X_1 X_2$	−0.0000	0.0000	−4.103	0.004
$X_1 X_3$	−0.0000	0.0000	−2.172	0.041
$X_2 X_3$	0.0001	0.0001	1.448	0.178

Table 5.8 ANOVA analysis for Eq. 5.14.

Source	Sum of squares	Degree of freedom	Mean sum of squares	F-value	$F_{0.05}$
Regression	0.318 75	9	0.0354	88.50	> 2.54
Residual	0.003 01	7	0.0004		
Total	0.321 75	16			
R^2	0.991				
Adjusted R^2	0.979				

yield of 1.12 at 5574 psi, 75 °C and 54 minutes. Optimum extraction yield is more than 1 which is not realistic. The yield cannot go beyond 1, but the operating conditions will be very close to optimal operating conditions.

5.9.3 Artificial Neural Network Modeling

As we discussed before, there is not a straightforward technique to determine the best training algorithm and number of hidden neurons. LM training algorithm was used at this research for data mapping. After several trial-an-errors, optimum number of neurons at hidden layer were determined as 50. Table 5.9 provides a summary of final network properties.

It will be interesting to compare ANN against the RSM model. The comparison has been provided at Table 5.10. As it is obvious from Table 5.10, ANN provides more accurate results. So the optimization using ANN will yield more accurate results. Basis of comparison is on root mean square error (RSM).

5.10 Subcritical Ethanol Extraction of Java Tea Oil

Orthosiphon (OS) which is known as "Misai kucing" in Malaysia is an herb that can be found in Malaysia, Indonesia, Thailand, and Philippines. The herb has several healing properties such as treating joint inflammation, arthritis, and rheumatism.

Java tea contains a number of phenolic compounds which play a significant role in the treatment of several diseases because of their anti-bacteria

Table 5.9 Parameters of the best ANN network during training process.

Training algorithm	Levenberg–Marquardt
Network	Feedforward
Hidden layer transfer function	tansig
Ouput layer transfer function	Purelin
Number of hidden layer neurons	24
Number of output layer neurons	1
Performance function	MSE
Divide function	Divider and
Best performance	6e−06
mu	1e−07
Gradient	2.73e−10

Table 5.10 Comparison of the ANN prediction with RSM for training data.

Temperature (°C)	Pressure (Psi)	Extraction time (min)	Oil yield (exp.)	Predicted value (ANN)	Predicted value (RSM)	Relative error (%) (ANN)	Relative error (%) (RSM)
40	5000	45	0.56	0.56	0.63	0.000 000	−12.500 00
40	7000	45	0.69	0.69	0.65	0.000 000	5.797 10
60	7000	45	0.89	0.89	0.90	0.000 000	−1.123 60
50	7000	30	0.64	0.64	0.63	0.000 000	1.562 50
50	5000	60	0.80	0.80	0.76	0.000 000	5.000 00
50	7000	60	0.74	0.74	0.77	0.000 000	−4.054 05
40	6000	30	0.63	0.63	0.60	0.000 000	4.761 90
60	6000	30	0.84	0.84	0.86	0.000 000	−2.380 95
50	6000	45	0.93	0.932	0.93	−0.215 05	0.000 00
50	6000	45	0.93	0.932	0.93	−0.215 05	0.000 00
50	6000	45	0.93	0.932	0.93	−0.215 05	0.000 00

properties. In the present study, OS oil has been extracted using subcritical fluid extraction.

Java tea was obtained from plants at Penang Island, Malaysia. The subcritical fluid extraction system is an accelerated solvent extraction system. This system has the advantage of fast extraction time in subcritical regions. The solvent for extraction is ethanol. Ethanol is safe for using in food applications and has good solubility for essential oils. When the extraction is finished, the extracted oil is collected in the collection cells, and they were placed in the evaporators for ethanol removal. When ethanol is evaporated then the oil samples were measured. The measurements and set of experiments were according to Box–Behnken experimental.

Oil recovery was the model output, while temperature, extraction time, and extraction cycle were selected as statistical model input parameters. Table 5.11 shows data level for scaling between −1 and 1.

In order to develop the RSM model, a second-order model like Eq. 5.21 was selected. The independent variables were scaled between −1 and 1 according to Eq. 5.24:

$$x_i = \frac{X_i - X_0}{\Delta X_i} \qquad (5.24)$$

where x_i, X_i, X_0, and ΔX_i are dimensionless coded value for the ith independent variable, uncoded value for the ith independent variable, unscaled value for the ith independent variable at the center point, and ΔX_i is the value of the step change, respectively. Result of DOE has been depicted in Table 5.12.

Table 5.11 Level of extraction process variables.

	Range and level			
Variables	Low level (−1)	Center level (0)	High level (+1)	ΔX_i^a
Temperature (°C)	80	100	120	20
Extraction time (min)	5	10	15	5
Extraction cycle	1	2	3	1

[a] Step change values.

5.10 Subcritical Ethanol Extraction of Java Tea Oil

Table 5.12 Box–Behnken design with coded/actual values and results.

Run	Coded level of variables			Actual level of variables			Observed result
	x_1	x_2	x_3	X_1 (°C)	X_2 (min)	X_3 (cycle)	Recovery (%)
1	−1	−1	0	80	5	2	7.129
2	1	0	−1	120	10	1	10.800
3	0	0	0	100	10	2	22.800
4	0	−1	−1	100	5	1	11.000
5	−1	1	0	80	15	2	10.601
6	−1	2	1	80	10	3	12.600
7	0	0	0	100	10	2	22.040
8	1	0	1	120	10	3	13.060
9	0	1	1	100	15	3	17.604
10	0	1	−1	100	15	1	11.550
11	1	−1	1	120	5	2	13.010
12	0	−1	1	100	5	3	14.690
13	1	1	0	120	15	2	10.324
14	−1	0	−1	80	10	1	9.0401
15	0	0	0	100	10	2	22.203
16	0	0	0	100	10	2	22.430
17	0	0	0	100	10	2	21.800

Equation 5.25 describes how temperatures (x_1), extraction time (x_2), and extraction cycle (x_3) are related to extraction yield.

$$Z = -217.085 + 3.816\,x_1 + 5.2703\,x_2 + 17.26\,x_3$$
$$- 0.0179\,x_1^2 - 0.193\,x_2^2 - 3.717\,x_3^2 - 0.016\,25\,x_1 x_3$$
$$- 0.015\,39\,x_1 x_2 + 0.118\,x_2 x_3 \tag{5.25}$$

ANOVA analysis was carried out to check the fitness of the model. As it has been shown in Table 5.13, Eq. 5.16 has R^2 equal to 0.9897 which confirms good fitting for the equation.

Table 5.13 Analysis of variance (ANOVA) of the response surface model.

Factors (coded)	SS[a]	df[b]	MSS[c]	F_{value} F_{cal}	Probability $(p) > F$
Model	423.7579	9	47.0842	68.4462	0.000 000[d]
X_1	7.6245	1	7.6245	11.0834	0.012 605
X_1^2	216.0512	1	216.0512	314.0640	0.000 000[d]
X_2	2.2578	1	2.2578	3.2821	0.112 941
X_2^2	98.1558	1	98.1558	142.6847	0.000 007[d]
X_3	30.2642	1	30.2642	43.9937	0.000 295[d]
X_3^2	58.1339	1	58.1339	84.5067	0.000 037[d]
X_1*X_2	9.4556	1	9.4556	13.7452	0.007 580[d]
X_1*X_3	0.4225	1	0.4225	0.6142	0.458 913
X_2*X_3	1.3924	1	1.3924	2.0241	0.197 822
E^e	4.8154	7	0.6879		
TOTAL SS	468.0075	16			

[a] Sum of squares.
[b] Degrees of freedom.
[c] Mean sum of squares.
[d] p values <0.05 were considered to be significant.
[e] E is indicating the error.

5.10.1 Artificial Neural Network Modeling

To develop ANN, 2/3 of experimental data were used for training ANN and 1/3 of the remaining data were used for generalization purposes.

Several training algorithms such as Conjugate gradient, Quasi–Newton, and Levenberg–Marquardt in the MATLAB environment were investigated. After exploring several training algorithms, "tansig" as a nonlinear transfer function and "purline" as a linear transfer function were selected for the final network. ANN network inputs were temperature, extraction time, and extraction cycles and output variable was extraction yield.

Table 5.14 compares different network estimation capability. As it is obvious from Table 5.14, Levenberg–Marquardt was found as the best training algorithms in this study. Optimum extraction condition based on RSM model were 100 °C, 10 minutes, and 2 cycles. The corresponding extraction yield for these

Table 5.14 Comparison of different ANN networks.

Observed data	7.1290	10.3240	14.6900	13.0100
Levenberg–Marquardt	7.1286	10.3250	14.6857	13.0104
Quasi–Newton	11.3778	14.1032	16.2320	12.8394
Conjugate gradient	8.2974	20.2625	11.1037	13.5550

values is 22.29%. The optimum extraction yield for ANN models is 22.8%. ANN model provides more extraction yield compared to RSM model. To inspect which yield is realistic accuracy of ANN and RSM models have been compared, value of the R_2 for RSM is 0.9897; while ANN model has R_2 value of 0.999. The result shows ANN model is more accurate than RSM model, so ANN higher yield is more trustable.

5.11 SFE of Oil from Passion Fruit Seed

Passion fruit seeds have oil which traditionally is extracted using filter press techniques. The seeds usually are distracted after use, but they have valuable oil. The oil is mainly used as anti-aging and for skin disease treatment. The oil also is known as Maracuja Oil.

At this study, SFE of maracuja oil using carbon dioxide has been modeled using RSM and ANN techniques. The study again provides good comparison between capabilities of these two techniques.

5.11.1 Experimental Procedures

Physical identification has been carried out for the seeds oil. Measurement of the density was done by the procedure of AOAC 985.19, while the color and the state of the oil were determined by visual check. The index of refractive index was identified at room temperature applying the Abbe refractometer (NAR-1T; Atago Co., Ltd., Tokyo, Japan). The plans for analysis of chemical indices, such as the acid value (AOAC 969.17), peroxide value (AOAC 965.33), saponification number (AOAC 920.160), nonsaponification matter (AOAC 933.08), iodine value (AOAC 993.20), insoluble impurities (AOCS Ca 3a-46), and moist and volatile matter (AOAC 926.12) were conducted through the official procedure. Each sample of oil was examined three times (Shucheng et al. 2009).

The determination of fatty acid profiles was implemented by preparing the methyl esters by the procedure of the AOAC 996.06. The examination of the fatty acid methyl esters was achieved by gas chromatography applying a Shimadzu GC-14B (Shimadzu Corporation, Tokyo, Japan), equipped with a flame ionization detector and integrator. A capillary column (30 m × 0.25 mm × 0.25 m), free fatty acid phase (FFAP) was purchased from the Dalian Institute of Chemical Physics (Dalian City, Liaoning Province, China), with a stationary phase of polyethylene glycol. The injector and detector temperature values were the same at 250 °C. The temperature of the oven was maintained at 190 °C for 15 minutes and then subjected to rise up to 230 °C with the rate of 5 °C/min and controlled at this temperature for the same period of time as the first step. The carrier gas was nitrogen at the pressure of 0.5 MPa. The determination of the fatty acid methyl esters was accomplished compared with standards and was valued by the area percentage of each fatty acid methyl ester. The analysis was completed three times.

5.11.2 RSM Statistical Modeling

In this study, the quadratic polynomial has been used to determine relation between input and output variables:

$$Y = \beta_o + \sum_{j=1}^{k} \beta_j x_j + \sum_{j=1}^{k} \beta_{jj} x_j^2 + \sum_{i<}^{k} \sum_{j=2} \beta_{ij} x_i x_j \tag{5.26}$$

where Y represents the dependent variables, β_o is the intercept term, β_j, β_{jj}, and β_{ij} are the linear, quadratic, and interactive coefficients, respectively. Table 5.15 depicted the variables for RSM modeling using CCD technique.

Process inputs are temperature (X_1), pressure (X_2), and extraction time (X_3), while the output is oil extraction yield (Y). Process inputs were scaled between −1 and 1 for statistical modeling. Table 5.16 provides details about coded process variables.

Statistica software has been utilized for experimental modeling. Table 5.17 shows the result of statistical modeling.

From Table 5.17 and by discarding nonsignificant terms, the statistical model which describes relation between processes variables are obtained as follows:

$$Y = -71.453 + 2.5348\,X_1 - 0.0218\,X_1^2 + 1.3856\,X_2 - 0.0266\,X_2^2 \\ + 4.276\,X_3 - 0.1635\,X_3^2 + 0.0021\,X_1 X_2 - 0.0385\,X_1 X_3 - 0.0323\,X_2 X_3 \tag{5.27}$$

Table 5.15 DOE for passion fruit seed oil extraction.

Run	Temperature (°C)	Pressure (MPa)	Extraction time (h)	Oil extraction yield (%)
1	60	30	5	24.7
2	60	30	1	24.39
3	60	20	5	24.4
4	60	20	1	23.14
5	50	30	5	24.19
6	50	30	1	22.68
7	50	20	5	24.44
8	50	20	1	21.3
9	63	25	3	24.27
10	47	25	3	23.78
11	55	33	3	24.37
12	55	17	3	23.06
13	55	25	6.36	24.75
14	55	25	0.36	23.4
15	55	25	3	25.39
16	55	25	3	25.6

Table 5.16 Levels of independent variables used in RSM design.

Coded-variable levels (Z_j)	Temperature (X_1, °C)	Pressure (X_2, MPa)	Extraction time (X_3, h)
+1.682	63	33	6.36
+1	60	30	5
0	55	25	3
−1	50	20	1
−1.682	47	17	0.36

Table 5.17 RSM model parameters and validation.

Factor	Regression coefficient	Std. error	t(6)	p	−95% Cnf. Lim.	+95% Cnf. Lim.
Mean/Interc.	−71.452	14.694	−4.862	0.0028	−107.410	−35.495
(1) Temperature (°C)(L)	2.534	0.461	5.498	0.0015	1.407	3.662
Temperature (°C)(Q)	−0.021	0.004	−5.361	0.0017	−0.032	−0.011
(2) Pressure (MPa)(L)	1.386	0.310	4.463	0.0042	0.626	2.146
Pressure (MPa) (Q)	−0.026	0.004	−6.552	0.0006	−0.037	−0.016
(3) Extraction time (h)(L)	4.276	0.661	6.460	0.0006	2.656	5.895
Extraction time (h)(Q)	−0.163	0.027	−5.969	0.0009	−0.230	−0.096
1L by 2L	0.002	0.004	0.497	0.6365	−0.008	0.012
1L by 3L	−0.038	0.010	−3.648	0.0107	−0.064	−0.012
2L by 3L	−0.032	0.010	−3.0566	0.0223	−0.058	−0.006

5.11.3 ANN Modeling of Passion Fruit Seed Oil Extraction with Supercritical Carbon Dioxide

Before using data, preprocessing on data was performed to eliminate off data. To develop ANN model, back-propagation learning with one hidden-layer network, Logistic Sigmoid, and purelin transfer and LM training algorithm were employed. Two-third of the data were used for model development, while one-third remaining data were employed for generalization purposes. By employing all the training algorithms, it was found that eleven neurons in the hidden layer provide best predictions (Table 5.18).

MSE values for ANN model and RSM method were calculated 0.0009 and 0.0421, respectively. Therefore, the results derived from the ANN model were significant and the error was incredibly low. Table 5.19 provides a good

Table 5.18 Parameters of the best network.

Training algorithm	trainlm
Number of neuron in hidden layer	11
Hidden layer transfer function	logsig
Output layer transfer function	purelin
Net training parameter (momentum constant) – mc	0.5
Net training parameter (Marquardt adjustment parameter – initial value) – mu	0.0610
Net training parameter (increase factor for mu) – mu_inc	8.6100
Net training parameter (decrease factor for mu) – mu_dec	0.6

Table 5.19 Comparison of the ANN with RSM for testing data.

Temperature (°C)	Pressure (MPa)	Extraction time (h)	Oil yield (exp.)	Predicted value (ANN)	Predicted value (RSM)	Div (%) (ANN)	Div (%) (RSM)
60	20	5	24.4	24.348	24.277	0.209	0.502
50	30	5	24.19	24.272	24.414	−0.340	−0.928
55	33	3	24.37	24.459	24.420	−0.367	−0.208
55	17	3	23.06	23.126	23.219	−0.289	−0.690
55	25	3	25.39	25.599	25.522	−0.826	−0.521

comparison between ANN and RSM models. The one-third test data have been selected to compare ANN prediction vs. RSM model. This was due to the fact to check network performance for unseen data. The test data were not used for developing ANN, while they were used for developing statistical models. So the first impression is that the RSM model should be more accurate than ANN model, but still ANN provides a better fit for the system.

As ANN is more accurate than RSM model, then the optimization results based on ANN model will be more accurate. Based on ANN model, the optimum temperature, pressure, and extraction time were found 56.50 °C,

23.30 MPa, 3.72 hours, respectively. The optimum oil extraction yield using ANN model is 26.55% compared to 25.76% using RSM model.

The comparison between ANN and RSM models has been demonstrated by MSE and relative error methods. For instance, MSE value for ANN model has been enumerated to be 0.0009 which is 0.0421 for RSM method.

As it was indicated at the beginning of this chapter, ANN models provide better results compared to RSM models if enough experimental data exist.

6

Experimental Design Concept and Notes on Sample Preparation and SFE Experiments

6.1 Introduction

The success of supercritical fluid extraction is influenced by many factors associated with process operations. The critical factors are pressure and temperature, whereas the indirect factors are solvent flowrate, extraction time, sample, and particle sizes as well as moisture content. Nevertheless, these factors have simultaneous effects toward the extraction studied and cannot be ignored in order to understand the actual behavior of the process. Therefore, the individual and interactive effects of operating conditions on the extraction yield are very important to be designed prior to experiments, and ultimately critical to determining the optimal extraction conditions. Experimental design is the best scientific tool to identify and optimize the significant factor. This chapter discusses the basic concepts of experimental design of supercritical extraction processes.

6.2 Experimental Design

Experimental design involves the process of planning a study to achieve research objectives and answer key research questions. A good planning prior to experimental works could save time, minimize unnecessary steps, and result in sufficient appropriate data being collected for further analysis. Experimental design can be classified into two categories, i.e. the screening and optimization designs. Under the screening category, the most frequently used designs include the Full Factorial design, Fractional Factorial design, and Plackett–Burman that are aimed at determining the significant factors among the influential factors, and their interactions (Brachet et al. 2000; Sharif et al.

Modeling, Simulation, and Optimization of Supercritical and Subcritical Fluid Extraction Processes, First Edition. Zainuddin A. Manan, Gholamreza Zahedi, and Ana Najwa Mustapa.
© 2022 by the American Institute of Chemical Engineers, Inc.
Published 2022 by John Wiley & Sons, Inc.

2014). On the other hand, Central Composite Design (CCD), Taguchi, and Box–Behnken are the most common optimization design tools (Sharif et al. 2014). These designs are however considered as more complex and called as second-degree design (Box et al. 1978). Numerous commercial design software such as Design Expert®, Statgraphic Centurion®, SixSigma, among others, offer all the available experimental designs tools for analysis.

6.3 Statistical Optimization

Traditionally, optimization of process conditions was simply carried out by conducting one factor at-a-time of the experiments. However, the results do not reflect the actual behavior of the process as interactions between factors that are present simultaneously are ignored. Besides, the one-factor optimizations usually comprises a large number of experimental works which leads to longer time and material consumptions.

To overcome the disadvantages of the traditional way, multivariate statistical analysis has been applied to examine relationships among multiple variables at a time. This technique analysis is performed when the combination of experimental factors has a different level (Ferreira et al. 2007). There are several multivariate statistical techniques employed in many research works, for example, multiple regression, path analysis, multivariate analysis of variance (MANOVA), and multivariate analysis of covariance (MANCOVA). However, it has been claimed that the most relevant multivariate technique in the statistical analysis is response surface methodology (RSM) (Bezerra et al. 2008).

In the application of SFE, RSM is the most popular and promising method to design an experimental work and to optimize SFE processes is by using RSM (Ghasemi et al. 2011; Li et al. 2010; Liu et al. 2009; Norulaini et al. 2009; Wei et al. 2009). In many SFE processes, RSM has been applied to investigate the individual or the interactive effects of operating variables. For example, SFE of oils from *Passiflora* (Liu et al. 2009) *Anastatica hierochunica* (Norulaini et al. 2009), sea buckthorn (Xu et al. 2008) grape (*Vitis labrusca* B.) peel (Ghafoor et al. 2010), and roselle seeds (Nyam et al. 2010) have been investigated using the RSM. On the other hand, Bhattacharjee et al. (2007) optimized the effects of temperature, pressure, and time of extraction of oil using statistical technique, i.e. central composite rotate design (CCRD) under RSM as well as to maximize the oil yield with minimum of gossypol extraction. On the other hand, Machmudah et al. (2007) applied a full 3^3 factorial design to optimize operation conditions of rose hip oil extraction using supercritical CO_2. The 27 experimental runs were performed randomly in determining possible interactions of process variables and their effects on rose hip oil and its compositions.

RSM is a statistical method that uses quantitative data from appropriate experimental designs to determine the relationship between test variables and response. The method also will simultaneously solve multivariate equations to represent the relationships. According to Giovanni (1983), the multivariate equations can be graphically represented as response surfaces which can be used in three ways: (i) to describe how the test variables affect the response; (ii) to determine the relationships among the test variables; and (iii) to describe the combined effect of all test variables on the response. The statistical model upon which the analysis of response surface designs is based expresses the response variable Y as a linear function of the experimental factors, interactions between the factors, quadratic terms. The second-order model is generally fit in RSM studies:

$$Y = \beta_o + \sum_{i=1}^{k} \beta_i X_i + \sum_{i \leq j}^{k} \beta_{ij} X_i X_j \qquad (6.1)$$

The three-level factorial is one of the types of designs for response optimization that are grouped under RSM. The design selected consists of all combinations of three levels of each experimental factor which uses only three levels of each factor, a low level (−), a central level (0), and a high level (+). Since there are only three levels for each factor, the appropriate model is a polynomial model which contains terms representing main effects, second-order interactions, and quadratic effects (Eq. 6.2):

$$Y = \beta_0 + \beta_1 X_1 + \beta_2 X_2 + \beta_{12} X_1 X_2 + \beta_{11} X_{12} + \beta_{22} X_{22} \qquad (6.2)$$

RSM has been increasingly applied for numerous applications especially for optimization purposes and was extensively reviewed in the literature. For example Garrigós et al. (2000) applied a full 2^4 factorial design for optimization of parameters for analysis of aromatics amines in finger-paints. They estimated the four factors (volume of modifier, static time, temperature, and pressure) on the aromatics amines recoveries by using two-level factor design.

There are several experimental design tools that can be utilized to execute the analysis of RSM. For the first-order or linear models, the most appropriate tool is a factorial design, whereas for second-order or quadratic models, CCD and Box–Behnken are the most relevant and suitable for the analysis and predictions (Bezerra et al. 2008; Xynos et al. 2014).

6.4 Optimization of Palm Oil Subcritical R134a Extraction

Palm oil is an important source of food and a major source of lipid. Steady increase in the world population increases the demand for palm oil as an important source of edible oils and fats. Almost 90% of the world's palm oil production is traded as edible oils and fats. Palm oil also accounts for about 13% of the total world production of oils and fats and is expected to overtake soybean oil as the most important vegetable oil (Sundram et al. 2003). The edible vegetable oil is conventionally extracted using the mechanical press process or solvent (Lau et al. 2008). In Malaysian palm oil milling, the common technique of extracting palm oil from fruits is by mechanical pressing using hydraulic press or screw press. The conventional processing of palm oil also leaves high content of carotenoids in pressed palm fibers. The pressed palm fibers found are a good source of carotene. The residual palm fibers from palm oil production contain between 4000 and 6000 ppm of carotenoids, six times higher than that found in crude palm oil (Franca and Meireles 1997). On the other hand, the current palm oil refining process has reduced the nutritive value of the palm oil. The tocopherol content was reduced and destroyed all carotenoids that are present in the palm oil.

As an alternative, numerous attempts have been reported on the application of greener and safer extraction techniques which may minimize the loss of valuable minor compounds in the palm oil called as supercritical fluids extraction (SFE). For over 10 years, only Bisunadan (1993) investigated the extraction of palm oil using CO_2 from palm mesocarp to examine the potential of SFE in extracting palm oil and simultaneously recover the valuable minor components. The experiments were done at pressures of 300, 400, and 500 bar with the temperature 40, 60, and 80 °C to investigate the effect of temperature, pressure, and sample pretreatment on the extraction yield as well as the comparison of the yield between extraction by $SC-CO_2$ and hexane. She found that the SFE method is more efficient in extracting carotenes, tocopherols, and triglycerides at high pressures and temperatures compared with hexane.

The latest study was by Lau et al. (2006) who investigated the characterization and supercritical CO_2 extraction of palm oil from palm mesocarp. The conditions surveyed were from 40 to 80 °C and 140–300 bar. They found that the minor components such as carotenes, vitamin E, phytosterols, and squalene were coextracted, and the overall quality of $SC-CO_2$ extracted crude palm oil appears equivalent to those obtained by commercial processing of crude palm oil (CPO).

For the recovery of valuable minor components such as carotenes, a number of studies have been done in recent years to overcome the recovery limitation. For example, attempts by Chuang and Brunner (2006) and Gast et al. (2001) involving deacidification, transesterification, and three steps of the SFE process to enrich the valuable carotenoids and tocochromanols have been published. These methods were developed since direct extraction of carotene using CO_2 was not very practical.

Extraction of palm oil from fleshy mesocarp using SFE technology is considerably still new, and to date, only a few studies have been reported on the extraction of palm oil by means of supercritical CO_2. Most researchers have worked on the extraction of oil from vegetables seed and reported on the solubility of the vegetables oils such as canola oil (Fattori et al. 1988; Temelli 1992), peppermint oil (Goto et al. 1993), rapeseed oil (Eggers et al. 2000) and, celery seed oil (Papamichail et al. 2000), sunflower oil (Kiriamiti et al. 2002) and olive oil (Hurtado-Benavides et al. 2004). Only a few of the studies worked on the extraction of palm oil from the fleshy mesocarp (Hurtado-Benavides et al. 2004; Lau et al. 2006). Latest Mustapa et al. (2009) has reported on the extraction of vegetable or seed oils from plant materials by method of SFE using 1,1,1,2-tetrafluoroethane (or R134a) in subcritical conditions as an alternative solvent to CO_2. The palm oil yield was increased with the increasing of pressure and temperature, and the yield was found as high as 66.06% w/w at 100 bar and 80 °C. The substantially high palm oil yield at relatively lower pressure exhibited that the subcritical R134a solvent is a viable alternative to supercritical CO_2 for the extraction of palm oil from its mesocarp. In other works, extraction of β-carotene from palm oil using subcritical R134a has shown the concentration of the extracted compounds, i.e. 330–780 ppm was comparable to the content found through a conventional processing method (Mustapa et al. 2011).

As a continuation for the extraction of palm oil using subcritical R134a, an optimization of the extraction process is necessary in order to have a better understanding on the behavior of the main parameters' effect on the extraction yield. The palm oil yield in subcritical R134a must be correlated as a function of the pressure and temperature. RSM was used to build a model to describe a relationship between palm oil yield and independent factors and to optimize the extraction conditions within the ranges by maximizing the palm oil yield.

The effects of process parameter temperature and pressure on the extraction yield of palm oil were investigated. CCD was employed to design the experiments and to optimize the oil yield. The two independent variables studied

Table 6.1 Treatment and coded levels of the independent variables for CCD.

Independent variables	Coded levels		
	−1.0	0	+1.0
A: Pressure (bar)	60	80	100
B: Temperature (°C)	40	60	80

namely temperature and pressure were set as low and high level. The temperature set used is 40 and 80 °C, whereas pressure is 60 and 100 bar. The coded and uncoded independent variables are shown in Table 6.1. The levels of the independent variables were based on the preliminary results. The analysis of RSM is based on expressing the response variable Y as a linear function of the experimental factors, interactions between the factors and quadratic terms (Giovanni 1983). The second-order model is generally fit in RSM as follows:

$$Y = \beta_0 + \sum_{i=1}^{k} \beta_i X_i + \sum_{i=1}^{k} \beta_{ii} X_{ii}^2 + \sum_{i \neq j}^{k} \beta_{ij} X_i X_j \qquad (6.3)$$

where Y represents the total oil yields in subcritical as g oil/g sample, β_0 is constant, β_i, β_{ii}, and β_{ij} are the linear, quadratic, and interaction coefficient, respectively, whereas X_i and X_{ij} are the coded level of the variables which influence the response variables.

6.4.1 Effect of Temperature and Pressure

The statistical analysis was applied to study the interactions between operating parameters such as pressure and temperature on the palm oil yield. The palm oil yields obtained from the experimental works according to the RSM design were listed in Table 6.2. The data were analyzed by multiple regressions to fit the quadratic equations to the dependent variables. The response surface and contour plots of the model was the best way to express the effect of any factor on the response within the experimental space. Response surface and contour plots for the estimated palm oil yield as a function of pressure and temperature are shown in Figure 6.1a and b, which represents the relation of the oil yield to the extraction pressure and temperature. As can be seen from Figure 6.1a, the palm oil yield increases with pressure and temperature. However, it is observed that the effect of increasing the extraction temperature is more

Table 6.2 Experimental design from RSM for the palm oil yield (g oil/g sample).

Standard order	Factor 1 A: temperature (°C)	Factor 2 B: pressure(bar)	Response: % yield (g oil/g sample)
10	0	−1.414	57.78
7	−1	+1	56.47
13	0	0	62.06
11	0	0	62.12
6	−1	−1	48.57
8	−1.414	0	52.16
5	+1	−1	57.83
1	0	1.414	63.92
3	0	0	62.00
9	0	0	62.09
4	+1	+1	66.06
12	0	0	61.98
2	1.414	0	63.33

significant than that for increasing the extraction pressure. For the temperature effect, further increase in temperature to 80 °C caused the oil yield to drop. This is because of the decrease of solvent power and density of the subcritical R134a with temperature. From the regression analysis, the equation obtained can be used to determine the maximum response, Y.

The interaction between temperature and pressure on the yield of palm oil extracted by subcritical R134a was also denoted in Figure 6.1 as a contour plot. All response plots showed clear peaks, implying that the optimum conditions for maximum values of the responses were attributed to temperature and pressure in the design space. This two-dimensional chart represented the responses on the temperature–pressure plane. The result clearly indicated that increasing the temperature leads to improvement in yield of palm oil in the pressure ranges investigated, but this effect was more significant at higher pressures.

Consequently, pressure had two opposite effects on the extraction process and the final trend with respect to pressure was dictated by the resultant of

(a)

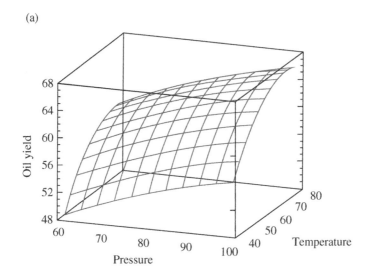

(b)

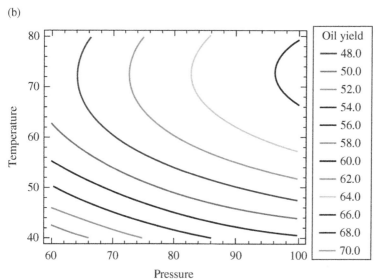

Figure 6.1 Extraction yield of palm oil as a function of pressure and temperature at 3 mL/min: (a) response surface (b) contour plots.

these two effects, which obviously depends on temperature level. The graph also shows the curvature produced on the yield by the temperature factor. Contour plots of the RSM were drawn in Figure 6.1b as a function of two factors at a time, holding all other factors at fixed levels (normally at the zero level). Those plots were helpful in understanding both main and the interaction effects of the two factors.

6.4.2 Model Fitting

From the regressions, the multiple coefficients were generated by employing a least squares technique to predict a second-order polynomial model, whereby the insignificant (P-value > 0.05) factors and interactions were eliminated from the model. The mathematical model established is very useful to describe the behavior of each response, summarized in Table 6.3. The accuracy of the model was evaluated from the coefficient of determination (R^2) and adjusted R^2. The coefficient of determination (R^2) is a ratio of the explained variation to the total variation. It is useful because it allows us to determine how certain one can be in making predictions from a model or graph. For the extraction yield of palm oil, the R^2 was 0.9913 and the adjusted R^2 was 0.9851 showed that the regression model defined the system behavior.

The models were found to be adequate for the data at the probability level of 99%. The significance of the independent variables on oil yield can also be observed from the Pareto chart (see Figure 6.2). It shows each of the estimated effects in the decreasing order of magnitude. The vertical line is used to judge which effects are statistically significant. Any bar that extends beyond the line corresponds to the effect that is statistically significant, at 95.0% confidence level. In this case, three effects are significant, i.e. individual temperature (B) and pressure (A) as well as interaction of temperature (BB). Therefore, the final empirical model obtained was in the form of Eq. 6.4, where the

Table 6.3 Regression coefficient for palm oil yield.

Coefficient	Value	p-Value
Constant	−16.5591	
A: Pressure	0.52045	0.0000
B: Temperature	1.4256	0.0000
AA	−0.002 170 26	0.0495
AB	0.000 206 25	0.7940
BB	−0.009 932 76	0.0000

Standard Error = 0.608298, R^2 = 99.13%, R^2 (adj) = 98.51%.
$Y = a_o + a_1 A + a_2 B + a_4 B^2$

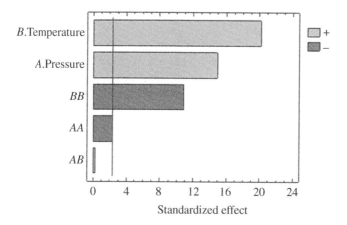

Figure 6.2 Pareto chart indicating the significant level for each parameter.

Table 6.4 ANOVA for quadratic model.

Source	F value	p-Value	Note
Model	23.62	0.0003	Significant
A-temperature	63.12	<0.0001	—
B-pressure	32.38	0.0007	
AB	0.011	0.9178	
A^2	20.84	0.0026	
B^2	3.65	0.0978	
Lack of fit	1583.38	<0.0001	Significant
Std. Dev	1.54	R^2	0.9441
Mean	59.72	Adj R^2	0.9041
Coefficient of variation, CV (%)	2.58	Pred R^2	0.6024
Press	118.25	Adequate precision	14.504

regression coefficients are also shown in Table 6.3. The quadratic equation was found to represent the experimental data well ($R^2 = 0.9913$ and AAD = 4.59%).

Analysis of variance (ANOVA) calculations were presented in Table 6.4. It shows that the regressions of palm oil extraction were statistically significant with $P < 0.0001$ and the models had significant lack of fit ($P < 0.05$). Therefore, the well-fitting model for palm oil extraction using subcritical R134a was successfully established. Furthermore, the way in evaluating the model is significant; the p-value must be less than 0.05. Therefore, from the table also note that temperature (A), pressure (B), and temperature-squared (A^2) were significant model terms. On the other hand, if the values are greater than 0.10, it indicated that the model terms were not significant, for instance temperature-by-pressure (AB) and pressure-squared (B^2). Therefore, these terms were removed from the model for further analysis. It is noteworthy that only the coefficient for interacting factor which had a value of 0.05 or less has been included in the model. However, if there were many insignificant model terms, model reduction may improve the desired model. In order to measure the optimal conditions to achieve the good extraction process of palm oil using subcritical R134a, Eq. 6.4 has been used to visualize the effect of factors on response.

$$Y = -16.5591 + 1.4256A + 0.52045B - 0.00993276A^2 \qquad (6.4)$$

where Y represents the palm oil yield, A and B are the coded variables for temperature and pressure, respectively.

6.4.3 Process Optimization

With multiple responses, the optimum condition where all parameters simultaneously met the desirable removal criteria could be visualized graphically by superimposing the contours of the response surfaces in an overlay plot. The main objective of this optimization was to determine the optimum values of temperature and pressure within their ranges by maximizing the palm oil yield. Increasing the pressure at a high temperature enhances the response (yield) as plotted in Figure 6.1a. Accordingly, it was also seen that minimum yield was predicted at a lower level of both pressure and temperature factors, whereas the maximum yield was predicted at higher level of pressure. However, there was an optimal value for the temperature to obtain the highest oil yield where lower or higher than this value of temperature could decrease the oil yield. From Figure 6.1a, it is observed that when the temperature lower than 73 °C, the yield was slightly decreased as pressure was increased, whereas when temperature above 73 °C, the yield steadily increased for all pressures range. On the other hand, with sufficient temperature, pressure of 100 bar was predicted to be the optimum pressure for the extraction at one with the entering data. Therefore, the optimum processing conditions to produce the maximum palm oil yield from treated palm fruit was at 100 bar and 72.80 °C with 66.43% yield. The value of yield was closest to the actual maximum yield point which was 66.06 g oil/g sample.

The optimization of palm oil using subcritical R134a showed that the temperature is a key factor that influences the yield rather than pressure effect. The second-order polynomial model applied was appropriate to describe and predict the palm oil yield within the experimental operating conditions ranges. The optimal conditions to obtain the highest palm oil were determined to be at 72.80 °C and 100 bar. Under these conditions, the palm oil yield was predicted to be 66.43% yield. The experimental values were found agreed with the predicted obtained from the modeling analysis. Therefore, it can be concluded that the optimization using this method is applicable in order to measure the optimal conditions for extraction of palm oil from palm mesocarp using supercritical fluid technology.

6.5 Comparison of Subcritical R134a and Supercritical CO_2 Extraction of Palm Oil

Extraction processes using SFE technology are potential alternative methods for vegetable and essential oil extraction to replace conventional industrial processes of expeller pressing, solvent extraction, and steam distillation. In SFE technology, supercritical CO_2 has been accepted as a common and standard supercritical solvent employed for most supercritical extraction applications. Using supercritical CO_2, high pressure of up to 500 bar is required to allow satisfactory extraction or fractionation process. The high-pressure operation can contribute to the high capital cost and operating cost to maintain the high pressure. Therefore, the discovery of a new or alternative low-pressure solvent having the same advantages as that of CO_2 is therefore necessary in order to capitalize on the superiority of SFE technology over traditional technique.

It is envisioned that the discovery of an alternative solvent that allows operations at significantly lower pressure relative to those required using CO_2 solvent is one of the alternative methods to overcome the high-pressure requirement by supercritical CO_2. Subcritical R134a is suggested as a low-pressure alternative (Catchpole and Proells 2001; Mustapa et al. 2009; Wood et al. 2002) to supercritical CO_2 since it has been found to have comparable solvent properties to CO_2 (Lagalante et al. 1998) in addition to being able to extract analytes at low temperature and pressure (Hansen et al. 2001; Mustapa et al. 2009; Simoes and Catchpole 2002).

For the solvent to be viable, it should enable the process to achieve comparable or better performance in terms of maximizing oil yields while maintaining the quality and stability of the oil through elimination of undesirable compounds. In achieving an economical and cleaner process, it is envisioned that R134a has a promising potential to replace CO_2 for the extraction of crude palm oil based on SFE technology. The performance of SFE is usually compared to other advanced and/or conventional extraction techniques such as ultrasound-assisted extraction (Sporring et al. 2005), microwave-assisted techniques (Lorenzo et al. 1999), Soxhlet extraction (Crespo and Yusty 2005), steam distillation (Kotnik et al. 2007) to investigate the successfulness of the SFE process over other techniques for the oil extraction.

For the same reason, it is important to determine the performance of supercritical CO_2 and subcritical R134a used in extraction of palm oil from its palm mesocarp. The key factors compared include the extraction performances and the economics of both solvents. The basis for comparison is effectively high yield and low cost since these are the ideal criterion for the best choice of solvent (Rivizi et al. 1986). Note that the cost of solvent and compression were considered when comparing and evaluating the economics.

6.5.1 Extraction Performance

A set of extraction was performed for supercritical CO_2 and subcritical R134a at 100 bar, 40 °C, 0.5 mL/min (Mustapa et al., 2011). This condition was chosen since 100 bar is the lower pressure limit for supercritical CO_2 processing (Brunner 1994), and the upper pressure limit for R134a. one of the key aspects in determining the capability of R134a, is to see whether it is comparable to SC-CO_2 for extraction at low pressures of CO_2. Besides, the combination of 100 bar and 40 °C was selected since the density of R134a is the highest compared to density at other combinations of temperature and pressure in this study. So that the solvating power of R134a is expected to enhance the extraction process. Solubility of β-carotene in subcritical R134a was also studied and compared with literature. Oil extraction yield is defined as the weight of extracted oil per unit sample used at given temperature and pressure in the extraction time. The oil yields were calculated using Eq. 6.5:

$$\text{Yield [g oil/g sample feed]} = \left(\frac{\text{extracted oil (g)}}{\text{sample feed (g)}}\right) \times 100 \qquad (6.5)$$

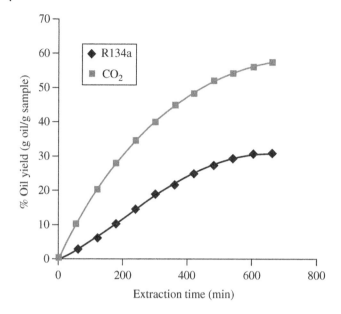

Figure 6.3 Comparison of extraction performance between supercritical CO_2 and subcritical R134a at 100 bar and 40 °C.

The results shown in Figure 6.3 shows that the extraction rate of CO_2 was much higher than the rate of extraction of R134a. In fact, the percent oil yield obtained using CO_2 was twofold higher than the yield obtained using R134a. It is clear that the extraction performance using CO_2 is superior to R134a even at the lowest CO_2 pressure. Even though the density of R134a is higher than that of CO_2, both solvents exist in two different regions. Carbon dioxide was operated in supercritical regions, whereas R134a was operated in subcritical or high-pressure liquid regions.

Operations at the two distinct regions contributed to different extraction characteristics. This situation was influenced by the mass transfer properties, including diffusivity and viscosity of solvents. Even though a material in the subcritical region was exploited in the same manner as a material in a supercritical region; however, the performance of the SFE process is dependent on the other related properties solvent such as diffusivity and viscosity when the extraction process involves solid sample.

To verify the state of both solvents, the reduced pressure and temperature for both solvents were calculated as the ratio of the actual absolute value to the critical point value. The reduced pressure and temperatures for CO_2 and

Table 6.5 Characteristics of CO_2 and R134a at several pressures and temperatures.

Solvents	P_r (bar)	T_r (°C)	Density (kg/m³)	Diffusivity	Viscosity (g/cm s)
100 bar-40 °C					
CO_2	1.372	1.288		$(2-7) \times 10^{-4}$	$^a(0.3-1) \times 10^{-3}$
R134a	2.463	0.396	1025.34	NA	$^b 1.963 \times 10^{-3}$
80 bar-60 °C					
CO_2	1.097	1.932		$(2-7) \times 10^{-4}$	$(0.3-1) \times 10^{-3}$
R134a	1.970	0.594	1120.32	N.A	$^b 1.510 \times 10^{-3}$

^{a} Given in Castro et al. (1994).
^{b} Calculated from method of Krauss et al. (1993) and Kiselev et al. (1999).
NA: Not available.
Sources: Adapted from Castro et al. (1994), Krauss et al. (1993), and Kiselev et al. (1999).
© John Wiley & Sons.

R134a are listed in Table 6.5. All reduced values of CO_2 are observed to be greater than unity. CO_2 is therefore in its supercritical state (Castro et al. 1994), while R134a is in its subcritical state (since the reduced temperatures of R134a are below unity).

The solubility of palm oil in subcritical R134a obtained also compared with the literature. Figure 6.4 shows the solubility of palm oil using CO_2 and R134a at temperature ranging from 40 to 80 °C and various pressures. Bisunadan (1993) carried out the extraction of palm oil from sonicated-dried mesocarp at a pressure range of from 300 to 500 bar. The reduced property was used to compare the solubility value since the ranges of pressures studied were different between Bisunadan's and this work. It is observed that low reduced pressure yielded low oil solubility.

The low reduced pressure of R134a solvent contributed to a very low solubility of palm oil compared to the case of supercritical CO_2 which led to a high solubility of palm oil. This may be due to the low solvating power of R134a as compared to that of CO_2 at the given conditions. This resulted in low solubility of palm oil in R134a solvent. The low-reduced pressure (hence solvating power and solubility) of R134a solvent cannot match the high solubility obtained by CO_2. This may be due to the solvents existing in markedly different regions, i.e. in subcritical regions for R134a and in supercritical regions for CO_2. Operation

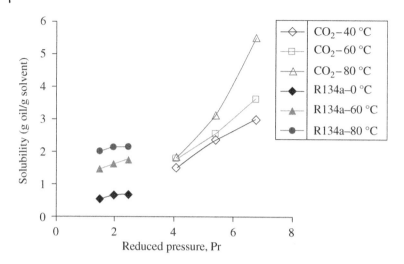

Figure 6.4 Comparison of palm oil solubility in terms of reduced pressure.

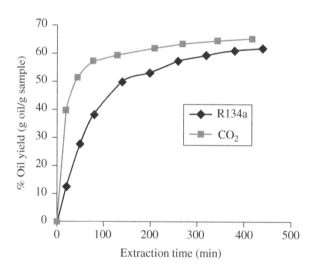

Figure 6.5 Comparison of extraction performance between supercritical CO_2 and subcritical R134a at 80 bar and 60 °C.

at different regions and solvents characteristics may yield markedly different performances.

Another extraction condition was carried out for CO_2 at 80 bar, 60 °C and 3 mL/min as shown in Figure 6.5. It was observed that the CO_2 extraction rate

Table 6.6 Costs comparison between CO_2 and R134a.

Solvent	Cost Material[a]	Cost Compression	Solvent used (g)[b]	Notes
CO_2	$16/cylinder (20 kg)	High pressure high operating cost	308.43	High pressure means high cost for compression High maintenance cost (Wood et al. 2002)
R134a	$45/cylinder (12 kg)	Low pressure low operating cost	1757.21	Recycle system help to reduce the cost of material (Catchpole and Proells 2001; Simoes and Catchpole 2002; Wood et al. 2002)

[a] In year 2021.
[b] Extraction at 80 bar and 60 °C in 440 minutes.

was higher than that of R134a. This occurred even when the applied pressure was below the low-pressure limit of CO_2. One reason for this performance was also due to the high diffusivity and low viscosity of CO_2 as compared to R134a. The higher R134a viscosity than supercritical CO_2 (refer to Table 6.5) had hindered the high mass transfer rates between solutes and R134a. This is because the high viscosity of R134a could have prevented solvent from penetrating the solid matrix efficiently and resulted in less oil being dissolved in R134a solvent. CO_2, on the other hand, has good mass transfer properties, including viscosity and diffusivity that are superior to that of R134a.

However, the results are different from those reported by Catchpole and Proells (2001). They found that lipids solubility in CO_2 at 60 bar and temperature of 40°–80 °C is effectively zero, whereas the lipids solubilities in R134a ranged from 0.81 g/kg R134a at 40 °C and 60 bar for soya oil to 9.89 g/kg R134a for squalene at 80 °C and 60 bar. The discrepancies may be due to different types of samples used in both studies. Catchpole and Proells (2001) used liquid samples of lipids and a countercurrent extraction process. Subcritical R134a solvent was fed downward, whereas lipids were pumped upward through the extraction column. The extraction process using a solid matrix was different from extraction using a liquid sample. This is because extraction of solutes from solid involves solute–matrix interactions, which was considered as one of constraints in solid extraction (Turner et al. 2001).

In order to enhance the extraction process that involves solid samples, good mass transfer properties of solvent can facilitate the removal of solutes from the matrix by two key steps, i.e. penetration of the solvent into matrix and dissolution of oil into the solvent through diffusion process. To promote the two steps efficiently, high diffusivity and low viscosity of solvent are required for the extraction process. Note that the properties are effectively applicable in the supercritical region of CO_2, rather than in the subcritical region of R134a. Thus, to operate the SFE process for solids, it is recommended that the supercritical conditions of CO_2 be used than R134a in its subcritical conditions. Otherwise, equal polarity of R134a solvent-substance can give a higher extraction yield than those achievable via CO_2 extraction since R134a is a superior polar solvent. Ashraf-Khorassani et al. (1997) showed that the solubility of polar solutes in subcritical R134a is much higher as compared to the solute's solubility in SC-CO_2 due to the higher polarity and/or density of R134a.

6.5.2 Economic Factor

SFE processes have been associated with high capital and operating costs. The cost for compression has been emphasized as one of the major factors that limit commercialization of SFE technology. Rivizi et al. (1986) reported that the extraction of fats and oils is effective only at very high pressures of up to 500 bar. Reverchon (1997) states that the cost of compression tends to complement the products when it has a high value as well as high-quality product.

In terms of material cost, carbon dioxide is cheaper than R134a and therefore has potential to reduce the material cost. Even though R134a is twice higher in terms of mass (g) and is a much more expensive fluid compared to CO_2 (refer Table 6.6); however, a recycle system is always considered as a necessity for industrial-scale operations due to economic and environmental reasons. Therefore, the cost of R134a has little influence on the economics (Catchpole and Proells 2001; Wood et al. 2002).

As a rule, high pressure means high capital and maintenance cost. R134a solvent has been investigated as an alternative low-pressure solvent for CO_2 for more than 10 years (Ashraf-Khorassani et al. 1997; Catchpole and Proells 2001; Simoes and Catchpole 2002; Wood et al. 2002). Substantial yields at relatively lower pressure (compared to extraction using CO_2) proved that subcritical R134a is a viable alternative solvent (Catchpole and Proells 2001). However, in this study, subcritical R134a does not demonstrate a superior and even comparable performance to CO_2. This was probably due to the different polarity between the R134a solvent and the substance (palm oil). Besides, CO_2 is also believed to have superior supercritical fluid properties compared to R134a solvent which existed in subcritical regions. However, it is believed that R134a solvent could be a great alternative solvent for CO_2 when the polarity of solvent-substance is equal.

Note that both solvents have unique advantages and disadvantages. For an effective and economical SFE process, first, it is recommended for the solvent to be compatible with the substance in terms of polarity. In addition, the products must also be high-value so that a reasonable profit margin can be maintained. R134a solvent could be a viable alternative to CO_2 if the solutes are polar and extraction do not involve solid materials. Finally, an efficient recycling system is a necessity.

All-in-all, the supercritical CO_2 is superior to that of R134a for extraction of oils from solid material. However, subcritical R134a is a better choice for extraction of oils at lower pressure than that of supercritical carbon dioxide. A detailed economic analysis should be performed to determine the production cost, total investment, variation and fixed cost, energy consumption, and profit by considering the yield of extraction, efficiency of the process, and prices of raw materials and equipment.

6.6 Sample Pretreatment

In a process of supercritical extraction, sample pretreatment plays an important role in enhancing the extractability of solutes from the solid matrices. Principally, alteration on the sample matrix could reduce the mass transfer resistance, change the direction of the extraction process (Vidović et al. 2014), improve penetration of solvent into the solid, and ultimately increase the extraction yield of compounds interest. In addition, optimum sample pretreatment also can minimize the solvent consumption and hence enables energy savings in the SFE process. Principally, there are several main characteristics considered in the sample pretreatment. This includes the sample moisture content and the particle size. A successful SFE process can be achieved when the moisture content and the particle sample size are optimized. The sample pretreatment can be done either by (i) simple pretreatment extraction such as microwave-assisted extraction (MAE) or ultrasonic extraction, and (ii) mechanical pretreatment such as grinding, chopping, crushing, or cutting.

6.6.1 Moisture Content Reduction

Successful SFE of seed oils not only requires optimized extraction conditions but also needs properly prepared and optimized seeds. For an optimized supercritical extraction, the influence of sample moisture content is important to consider during sample preparation and experiment. First, the moisture content affects the extent of sample heterogeneity. Second, it affects the experimental procedure in the sense that, when a fresh sample is extracted, its high moisture content can cause mechanical difficulties such as restrictor clogging due to ice formation (Lang and Wai 2001). Third, high moisture content can inhibit contact between fluid and lipid-rich region and lead to low recovery (King 1997) since highly water-soluble solutes would prefer to dissolve into the aqueous phase (Weathers et al. 1999).

In addition, several studies have shown that reduction of water content prior to the SFE process has a significant effect on the extraction yield. Bulley et al. (1984) stated that the efficiency of oil extraction also depends on the particle size and moisture content apart from pressure, temperature, and contact time. King (1997) emphasized that the kinetics of SFE are not only dependent on the extraction pressure and temperature but also on seed preparation and its morphology. Eggers (1996) states that the extraction process efficiency is directly related to the pretreatment of feed materials.

In general, excess water content in fresh samples should be removed to prevent any problems during extraction and to achieve effective SFE performance. The excess moisture content can be removed using one of several methods including vacuum oven, lyophilization or freeze-drying, sun-drying, oven-drying, and microwave treatment. In the latter method, radiation of the microwave toward the water content in the sample elevates its temperature thus decreasing the moisture of the samples (Uquiche et al. 2008). A general rule of thumb for sample drying is that it should be done as rapidly, and at as low a temperature as possible (Curren and King 1997) to minimize the loss of temperature-sensible compounds.

There are limits for sample moisture content that are acceptable for an extraction. However, certain levels of moisture may be required for prolonged food storage to prevent oxidative deterioration when moisture levels are too low. For example, vegetables such as carrots and potatoes will develop oxidized flavors or become rancid in two or three weeks at a 2 or 3% moisture content. Otherwise, the oxidative deterioration of these foods is inhibited for several months when they have 8–10% moisture content (Curren and King 1997).

On the other hand, for the SFE process, the sample moisture content must be as low as possible to be acceptable. For the extraction of high valuable compounds from herbs, the sample moisture content of 3.0–4.0% was preferable (Senorans et al. 2000; Sun et al. 2002; Yepez et al. 2002). Snyder et al. (1984) reported that the acceptable sample moisture content might range from 3 to 12% to avoid any adverse effect on oil extraction. Nevertheless, the allowable moisture content in plants for the SFE process is relying on the chemical nature of active compounds of the herbal drugs (Vidović et al. 2014). If the moisture content of the substrate is in between 3 and 12%, the influence of the moisture on the mass transfer and solubility is negligible (Goodrum and Kilgo 1987). On the other hand, higher extraction yield can be achieved if the moisture content is high up to 25%. This case is particularly for decaffeination of coffee and nicotine from tobacco since the compounds are strongly bound to the plant matrix. Thus, high water level content is needed in order to enhance the release of the solutes from the plant (Balachandran et al. 2006; Hubert and Vitzthum 1978; Zosel 1974).

Note that the presence of turbidity oil indicates the presence of water in the extract. However, water can be removed quite readily by the addition of a filter aid or other absorbent (King 1997). Water content in samples also lead to hydrolysis and degradation of the analytes due to increasing acidity in the water caused by absorption of CO_2 (You et al. 1999). Nevertheless, in some cases, moisture may benefit SFE since it could facilitate extraction by acting as a modifier for fluids (Hawthorne and King 1999; Lang and Wai 2001). Moisture can also help preserve food quality by maintaining food freshness.

6.6.2 Sample Size Reduction

Sample properties such as the nature of the sample matrix, the moisture content, and sizes are important prior to the extraction process. These factors will affect the design and planning of experimental conditions, including the extraction duration and the flowrate to be used (Castro et al. 1994).

Basically, a large amount of sample will increase the fluid volume consumed and subsequently require longer extraction time. When the sample amount is further increased, this may result in lower solute recovery. However, once the extraction conditions are optimized for larger samples, the results may be similar to that of smaller samples. In general, sample volume and extraction time may strongly affect the extraction course when the solute concentration in the sample is high. Besides, the extraction rate increases with decreasing particle size (Brunner 1994; Castro et al. 1994; Lang and Wai 2001; Reverchon 1997) and with increasing sample surface area to weight ratio (Castro et al. 1994).

Instead of grinding the plant sample into smaller particles sizes, other several effects of mechanical treatment such as flaking, cutting, impact plus shearing, and chopping also have been investigated by several researchers (Ivanovic et al. 2014; Mustapa et al. 2009). It has been found that chopping of palm fruits samples had contributed to the highest yield in comparison to the yield obtained by samples that flaked and cooked (Mustapa et al. 2009). However, depending on plant materials, different behaviors can be observed. Ivanovic et al. (2014) reported that flaking was found as the most effective pretreatment for SFE of fennel seeds, artemisia aerial parts, and the lichen *Usnea barbata*, whereas the technique of impact plus shearing was the optimal pretreatment for amaranth, milk thistle, and oat seed prior to the SFE process. All in all the possible reasons that might explain the effect of the mechanical on extraction yields are types of plant materials, oil contents, and location of solutes in plant materials. Apart from the mechanical techniques, microwave treatment has been reported to change granules seeds to powdery seeds which result in higher yield due to the increases of contact surface (Da Porto et al. 2016).

Nevertheless, in certain cases, particle size had no influence on the extraction of essential oils. Coelho et al. (2003) had reported that for different particle sizes, no change in the yield of essential oil from fennel fruit (Apiaceae) was observed. The same finding has been reported by Bocevska and Sovová (2007), where the yield of essential oil from yarrow flowers that were fine milled and cut by scissors was dependent on extraction time and temperature but not by particle size of the plant. In the case of plants that have secretory ducts, for instance Marigold and chamomile, it has been demonstrated that the particle size does not affect the extraction rate particularly for fine milled and cut particles to length of 5 mm (Zizovic et al. 2007). Thus, Al-Otoom et al. (2014) in their work on the extraction of olive oil have demonstrated that it is possible to perform supercritical extraction without size reduction which could reduce the cost of operation and increase the separation propensity between solid and the extracted oil. The maximum yield can be achieved as high as 12.3 wt% with the extraction time as the main factor that affects the extraction performance, followed by temperature and finally the extraction pressure.

6.7 New Trends in Pretreatment

Several new approaches have been introduced in the sample pretreatment prior to the SFE process. Microwave-assisted extraction (MAE), ultrasonic and enzymatically assisted extraction are among the common techniques being employed in the pretreatment and are considered as environmentally clean technology, time-saving, and significantly improved the extraction yield.

Mariano et al. (2009) carried out an enzymatic extraction prior to the hydraulic pressing of Pequi fruits (*Caryocar brasiliense*) and found that the extraction yield was increased at least 20% in comparison to the sample without enzymatic treatment. On the other hand, Passos et al. (2009) also has shown that the yield of grape seed oil that was enzymatically pretreated was significantly 44% higher than the yield obtained with an untreated sample. In the pretreatment process, the cell wall degrading enzyme cocktail of cellulose, protease, xylanase, and pectinase has been used to enlarge the broken/intact cell ratio, which hence increases the solute's extractability.

Numerous reported studies have shown that samples that are pretreated with microwave-assisted extraction (MAE) and followed with SFE process contributed to a significantly higher extraction yield compared to yield with SFE alone. Microwave radiation applied as a preheating technique that destroys the plant cell walls and provides larger accessible pores for solvent and solutes release before the extraction of oil from the seeds or plants. The situation begins when the solvent and moisture within the plant material absorbs the microwave energy, results in evaporations and generates tremendous pressure inside the cell walls. This internal pressure causes the cell walls to burst and destroy the plant material structure (Tatke and Jaiswal 2011). Microwave-assisted pretreatment has received increasing attention over the last several years due to numerous advantages, e.g. lower cost, shorter time, higher extraction kinetics, and less or without organic solvent.

There are several methods of microwave pretreatment has been employed such as microwave-assisted extraction (MAE), pressurized-microwave-assisted extraction (P-MAE), or solvent-free microwave extraction (SFME). In the MAE and P-MAE techniques, a mixture of water-organic solvent is used as a medium to absorb energy from the microwave radiation and dissipate the heat to the system. The selection of the solvent is mainly dependent on the selectivity with the compounds of interest and the dielectric constant and loss factor of the solvent. High dissipation factor values indicate high capability of the solvent to absorb the microwave energy (Veggi et al. 2013). Small volume of the solvent employed could accelerates the treatment process and promote better cell walls rupture. Numerous studies had demonstrated that microwave pretreatment on plant material prior to application of SFE method has successfully increased the oil extraction yield. For example Durđević et al. (2017) reported that microwave pretreatment on the pomegranate seed at 250 W for 6 minutes had contributed to the highest yield of 27.2% without destroying the composition of fatty acids of the seed oil.

In principle, sample pretreatment using P-MAE leads the plant cell walls rupture severely due to the presence of a pressure system that enhances the disruption of the plant material structure. Nevertheless, in several cases, P-MAE solely improves the extraction without any variations in the range of compounds extracted (Mustapa et al. 2015). On the other hand, in the SFME technique, the treatment is carried out without any solvent or water. The moisture content in the plant material itself acts as a medium for the energy absorption and leads to the plant swelling thus bursting due to high heat absorbed in the internal plant material (Li et al. 2013; Tatke and Jaiswal 2011).

Studies have reported that a microwave pretreatment on the plant samples prior to the extraction of oil has improved the yield of extraction and the quality of oil. For example, Dejoye et al. (2011) obtained between 25 and 150% higher extraction yield of microalgae oil from *Chlorella vulgaris* than the yield from SFE without the MAE treatment. In fact, the oil was apparently high quality and rich in fatty acids especially saturated polyunsaturated fatty acids. In addition, by treating the samples with MAE followed by SFE, there was severe damage to the plant cell matrix. In other cases, MAE also has been applied as an air-drying method prior to the SFE process. By using the combination of MAE and SFE process, the oil recovery has been achieved as high as 95% in less than 20 hours by supercritical CO_2 at 40 °C and 20 MPa (Quitain et al. 2013). The technique was more efficient than the conventional cold-press method, which usually had oil recovery in between 30 and 50% and longer extraction time from 48 to 72 hours. Nevertheless, to some extent, increasing the microwave radiation time pretreatment on seeds materials before the SFE process decreased the yield of the extraction. Longer the radiation time causes the particles size of the seeds materials to be extensively reduced and give high porosity of the SFE extraction bed (Da Porto et al. 2016). This problem could lead to bed compaction, irregular flow of $SC-CO_2$, and pressure channeling. When this happens, the contact between the solvent and solutes will be reduced and result in the decrease of the extraction yield.

Ultrasonic-assisted extract (UAE) also has been applied prior to the SFE process in order to enhance the extractability of compounds from solid matrix. Principally, ultrasound waves cause vibration to occur in the interfaces between solvent and solid, which leads to the damage of cell walls and eventually increases the solute's releases to the extraction medium. The disruption of plant cells is due to the collapse of cavitation bubbles from the ultrasonic wave near the cell walls. In addition, it also has been reported that ultrasonic is likely to increase the number of ruptured cells and provide faster access for the solvent to extract the solutes (Balachandran et al. 2006; Santos et al. 2015;

Toma et al. 2001). With the aid of UAE, the extraction yield was significantly increased as high as 14% compared to the yield without ultrasonic treatment and with less-severe operating conditions (Hu et al. 2007). Liu et al. (2020) reported that, by employing UAE treatment on samples prior to SFE has enhanced the extraction yield to between 11 and 28% higher as compared to the yields achieved without ultrasonic treatment. In addition, variation of fatty acids and phytochemical compounds were also found to have increased, clearly indicating an improvement in the quality of the oil when the plant was subjected to ultrasonic treatment of between 5 and 10 minutes at 0.625, 1.25, and 2.5 W/m of ultrasonic energy density.

Nevertheless, in some cases, despite the extraction rate has been improved, ultrasonic waves did not influence the extraction of several compounds of interest. Santos et al. (2015) claimed that the ultrasonic solely enhanced extraction rate and did not affect the extractability of phenolic compounds and capsaicinoid. This happened due to the nonruptured of cell walls observed on smaller particles. However, the effect of ultrasonic was found to be significant on larger particle sizes.

6.8 Optimal Pretreatment

Zizovic et al. (2007) has demonstrated that an optimal pretreatment beyond solely grinding has significantly enhanced the extractability of solutes prior to the supercritical extraction. The optimal pretreatment has been characterized by swelling of plant material which leads to increased diffusion coefficient that causes faster mass transfer of solutes. The swelling of plant materials might occur in two conditions: (i) during the fast decompression of SC-CO_2 which in turn caused a significant tensions created inside the plant cell, (ii) during the exposure at high pressure which the high dense CO_2 diffuses into the plant cells, dissolving the oils, expanding the plant material and causing the swell (Stamenic et al. 2010). In the optimal pretreatment, the materials-solvent are left to stand in an hour in supercritical condition before the extraction. As a result, the yield of extraction was increased and contributed to the large energy savings by minimizing the amount of solvent consumed to solubilized 1 g of extract. The advantages of the optimal pretreatment are that the consumption of carbon dioxide can be reduced if the SFE process commences from the swollen plant materials (Stamenic et al. 2010; Zizovic et al. 2007).

7

Supercritical and Subcritical Optimization

Part I: First Principle Optimization

7.1 Introduction

Simulation and optimization methodologies are widely used for process design studies in the chemical industry to reduce the experimental work required. The use of computer simulation techniques is particularly advantageous for the design of SFE processes, since it allows a broad range of operating conditions and process configurations to be explored quickly and easily. Using the developed RKA thermodynamic model and the proposed steady-state equilibrium-stage extractor model in Chapter 6, different SFE schemes were applied to explore the feasible operating domains for the production of refined quality palm oil using supercritical CO_2. In the following section, a systematic study to define the optimal conditions of temperature, pressure, S/F ratio, and reflux ratio was carried out in order to achieve the goal of improving the product quality while maximizing the yield of refined palm oil at the same time.

7.2 Evaluation of Separation Performance

Once the feasible operating conditions phase equilibria were determined, preliminary evaluation of separation process parameters such as number of stages, S/F ratio, product purity, and yield was performed using the Aspen Plus® 10.2.1 process simulator. Separation analysis generally resulted in a preliminary estimation for optimum operating condition taking into account that a higher solvent density often yields in a higher loading and a decreased selectivity.

Modeling, Simulation, and Optimization of Supercritical and Subcritical Fluid Extraction Processes,
First Edition. Zainuddin A. Manan, Gholamreza Zahedi, and Ana Najwa Mustapa.
© 2022 by the American Institute of Chemical Engineers, Inc.
Published 2022 by John Wiley & Sons, Inc.

7.2.1 Effects of Temperature and Pressure

The equilibrium solubility of CPO in supercritical CO_2 and the separation factor between FFA and palm oil TG at different temperatures and pressures were considered. The feasible pressure–temperature range for the countercurrent SFE process of CPO-supercritical CO_2 system is shown by the shaded area in Figure 7.1.

In this study, extraction temperatures below 370 K were chosen to avoid thermal degradation of palm oil minor components that can occur at temperatures higher than 370 K. The extractor pressure was kept below 30 MPa to remain within the two-phase region in the temperature range mentioned. The operating conditions for the deacidification of the CPO were set at 20–30 MPa and 373 K to allow a reasonable density difference to exist between the coexisting phases so that two-phase countercurrent flow can occur in the packed column. The proposed SFE process was rigorously simulated and studied for the operating condition of 370 K and 20–30 MPa.

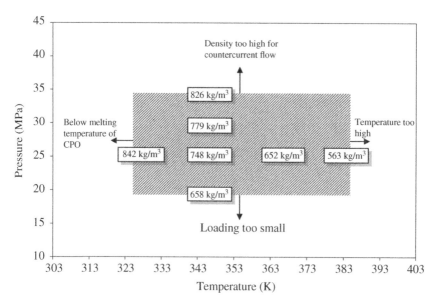

Figure 7.1 Investigated pressure–temperature range for countercurrent SFE process of CPO-CO_2 system (density of pure CO_2 given in box).

7.2.2 Effect of the Number of Stages

The removal of FFA from CPO in a continuous countercurrent extraction column can be enhanced by increasing the number of equilibrium stages as well as by performing multistage operations. For operation with multiple stages, feed was introduced at the column top (Stage 1). In this case, the influence of the number of stages on the process parameters for the production of 0.1 wt% FFA palm oil was considered. Figure 7.2 shows the effect of the number of stages on the separation behavior without external reflux at 373 K and 20–30 MPa. For a given feed specification, the separation performance of CPO deacidification process was improved with increase in the number of stages. As the number of stages increases, the contacting time and contacting area between the coexisting phases of the CO_2-palm oil system are increased. Due to this reason, the mixture separability and, hence, the purity of FFA in the extract phase of the extraction column increases with the number of stages.

Figure 7.3 shows the effect of equilibrium stage on the product recovery in the raffinate phase. The recovery of refined palm oil was found to gradually increase with an increase in the number of stages. Consequently, the removal

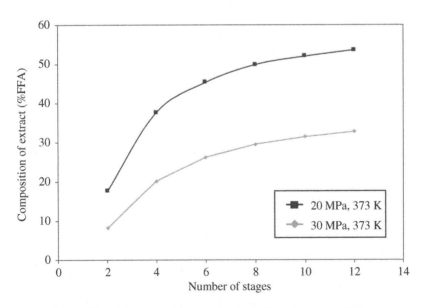

Figure 7.2 Effect of the number of stages on the recovery of FFA.

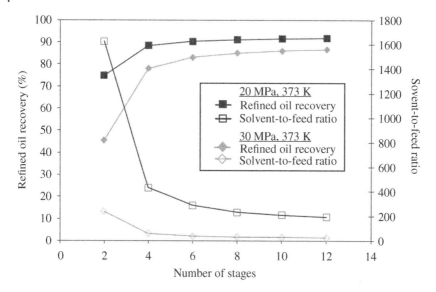

Figure 7.3 Effect of the number of stages on the recovery of refined palm oil.

of FFA was achieved without decrease in oil recovery. It can be concluded that the increase in the number of stages improves the mixture separability subject to the economic constraint as the optimum number of stages is a trade-off between the column capital and the operating cost.

7.2.3 Effect of Solvent-to-Feed Ratio

The flow rate of supercritical CO_2 solvent is an important factor to design the extraction process. Countercurrent extraction for the production of refined palm oil was performed using a 12-stage column without external reflux. For each operation, CPO feed was introduced at the column top (Stage 1).

Figure 7.4 shows the effect of S/F flow ratio on the FFA content of the extract phase and raffinate phase evaluated at 370 K and 20–30 MPa. At lower S/F flow ratio, higher FFA content was obtained in the extract due to higher solubility of fatty acids in supercritical CO_2. In contrast, as S/F value increases, the TG content in the extract increases as more of the feed dissolves in the CO_2-rich supercritical fluid phase. Simultaneously, the raffinate phase also becomes richer in TG as the S/F ratio increases due to an increase in the selectivity of the supercritical CO_2 solvent as S/F flow ratio increases. According to Ruivo et al. (2001), the variation of solvent and feed stream flow rates with S/F

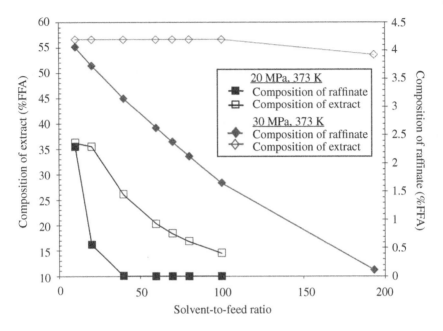

Figure 7.4 Effect of S/F ratio on the %FFA in the extract and raffinate.

ratio kept constant do not significantly affect the compositions of the extract and raffinate phases.

Figure 7.5 shows that an increase in S/F flow ratio causes a decrease in the recovery of palm oil in the raffinate phase. As the S/F flow ratio increases, some of the TG initially present in the feed is dissolved in the supercritical fluid phase, thereby resulting in the mass ratio between the raffinate and extract flow to decrease. The product quality increases with an increase in the S/F ratio, whereas the recovery of product decreases. For a given calculation, it can be concluded that effective separation is achieved by removal of FFA from the bottom product using higher S/F ratio.

7.2.4 Effect of Reflux Ratio

The effect of reflux flow on product recovery for countercurrent extraction of CPO mixtures was investigated at 373 K and 30 MPa. The S/F flow ratio and a range of reflux pump-rates for a 12-stage column were fixed. For each operation, the reflux stream was introduced at the column top (Stage 1). Figure 7.6 shows the effect of reflux ratio on the composition of the product (refined palm

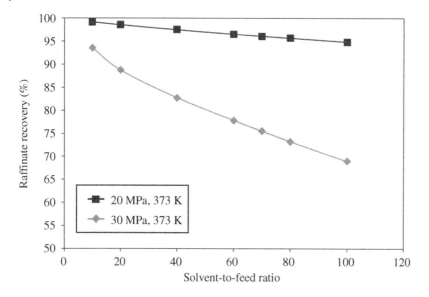

Figure 7.5 Effect of S/F ratio on the raffinate recovery.

oil) for CPO mixture at 373 K and 30 MPa with S/F ratio of 40. Figure 7.6 shows that the FFA content in the refined palm oil and top product increases with an increase in the reflux ratio. These results show that refluxing the top product is an effective technique not only for the fractionation of CPO but also for the purpose of concentrating highly volatile components such as FFA. Figure 7.7 shows the effect of reflux ratio on the recovery of refined palm oil in the raffinate phase. Recovery of refined palm oil increases with an increase in the reflux ratio and decreases with an increase in S/F ratio.

7.3 Parameter Optimization of Palm Oil Deacidification Process

The primary function of the extraction column is to produce a refined quality palm oil with FFA content of less than 0.1 wt%. In this study, the deacidification of CPO with supercritical CO_2 was studied as a mathematical programming problem. The extractor with 12 equilibrium stages was used. The CPO feed with FFA content of 4.5 wt% was fed to the stage-6 of the extractor. Different extraction schemes were studied, and optimal operating conditions were determined. Process alternatives considered for detailed study include (i) simple countercurrent extraction; (ii) countercurrent extraction with external reflux.

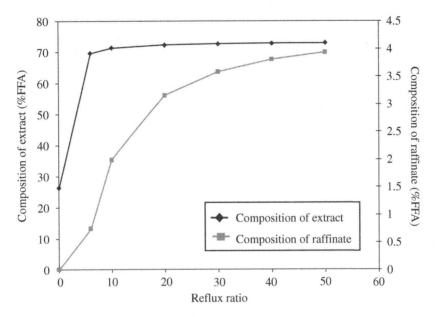

Figure 7.6 Effect of reflux ratio on the composition of the top and bottom products at 373 K and 30 MPa at S/F ratio of 40.

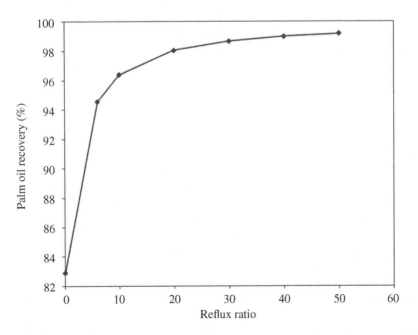

Figure 7.7 Effect of reflux ratio on the recovery of palm oil (S/F ratio of 40).

Table 7.1 Nonlinear inequality constraints for SFE palm oil refining.

Equipment	Limitation	Specification (%)
Extractor	FFA in raffinate (CO_2-free basis)	$\leq 0.10 \pm 0.1$
Extractor	TG recovery	$\geq 85.0 \pm 0.1$
Separator	FFA in extract (CO_2-free basis)	$\geq 85.0 \pm 0.1$
Separator	CO_2 in recycle stream	$\geq 99.99 \pm 0.1$

Optimization is an important analysis to achieve a specified objective function by manipulating decision variables associated with some specified constraints (Aspen Tech 2000). The optimization problem for the determination of optimal operating conditions has been formulated as a nonlinear-programming model. The Successive Quadratic Programming (SQP) (Biegler and Cuthrell 1985) optimization algorithm implemented in the Aspen Plus® process simulator was utilized to solve the model. Table 7.1 shows the specified inequality constraints for SFE process parameter optimization.

The variables that have the greatest impact on the process cost and performance are the extractor pressure and temperature, the solvent-to-feed ratio, the number of stages in the extractor, the reflux ratio in the extractor, and the separator pressure. In this study, several of these variables were fixed to make optimization more tractable. The number of stages in the extractor was specified because their inclusion in the design variables would require the solution of a mixed-integer, nonlinear program (with the number of stages as integer variables), which would significantly increase the computational complexity of the problem. The Newton-based method that solves the MESH equations is extremely sensitive to the initialization, and convergence failures are common. For this reason, the temperature of the extractor is fixed, and the optimal design is thus located at discrete temperatures.

In this study, two alternative objective functions were considered: (i) maximum palm oil TG recovery, subject to 0.1 wt% FFA concentration in raffinate, in a CO_2-free basis; (ii) minimum solvent recirculation.

7.3.1 Simple Countercurrent Extraction (Without Reflux)

This is the simplest extraction scheme with low product purity and recovery. Table 7.2 shows bounds for the optimization variables for SFE palm oil refining. Process simulation was performed at 373 K and 20–30 MPa. Table 7.3 gives the optimum values of the main process variables. These numerical results

7.3 Parameter Optimization of Palm Oil Deacidification Process

Table 7.2 Bounds for optimization variables.

Process variable	Lower bound	Upper bound
Extractor temperature (K)	333	373
Extractor pressure (MPa)	20	30
Separator pressure (MPa)	10	20
Solvent-to-feed ratio	10	100

Table 7.3 Optimum operating conditions for simple countercurrent process.

Process parameter	NLP optimum	
Objective function	Maximum TG recovery	Minimum CO_2 recirculation
Extractor temperature (K)	373	373
Extractor pressure (MPa)	20	30
Separator pressure (MPa)	11.5	11.5
Solvent flow rate (kg/h)	5245	1080
S/F ratio	87.42	19.46
FFA in extract (wt%)	48.92	31.91
FFA in raffinate (wt%)	0.1	0.1
CO_2 in recycle (wt%)	99.99	99.99
TG recovery (%)	91.75	86.53

show that a rather high purity palm oil (with 0.1 wt% FFA, in a CO_2 free basis) can be obtained with slightly lower TG recovery of ~90%.

7.3.2 Countercurrent Extraction with Reflux

In this scheme, the extract is depressurized and part of the liquid is recycled from the separator into the extraction column as a reflux stream. The reflux ratio is defined as the flowrate of the top product that is redirected into the column divided by the flowrate of the extract product. As the supercritical CO_2 flows through the extraction column, it selectively dissolves the FFA, leaving the TG in the raffinate. An increase in solvent flow rate improves the separation of the more volatile component (FFA and tocopherols). This, however, corresponds to a decrease in palm oil TG recovery (as shown in previous section). The existence of an external reflux increases the liquid flow rate in the extraction column, with a consequent increase in palm oil TG recovery. Optimization variables and their bounds are shown in Table 7.4. Table 7.5 shows the numerical results for optimum operating conditions at minimum solvent recirculation.

Table 7.4 Bounds for optimization variables.

Variable	Lower bound	Upper bound
Extractor temperature (K)	333	373
Extractor pressure (MPa)	20	30
Separator pressure (MPa)	10	20
Solvent-to-feed ratio	10	100
Reflux ratio	10	100

Table 7.5 Optimum operating conditions for countercurrent SFE with external reflux.

Process parameter	NLP optimum	
Objective function	Maximum TG recovery	Minimum CO_2 recirculation
Extractor temperature (K)	373	373
Extractor pressure (MPa)	20	30
Separator pressure (MPa)	11.5	11.5
Solvent flow rate (kg/h)	6545	3220
S/F ratio	109.08	53.67
Reflux ratio	8	30
FFA in extract (wt%)	85.97	86.13
FFA in raffinate (wt%)	0.1	0.1
CO_2 in recycle (wt%)	99.99	99.99
TG recovery (%)	95.01	94.95

7.4 Proposed Flowsheet for Palm Oil Refining Process

Summarizing the results of the previous section, the basic structure of the SFE process for palm oil refining can be outlined. A proposed flowsheet for the deacidification of crude palm oil, annotated with typical operating conditions, is shown in Figure 7.8.

The extractor was modeled with 12 equilibrium stages, with CPO entering continuously at the top and supercritical CO_2 at the bottom of the extractor. As the CO_2 flows through the extractor, it preferentially dissolves the high volatile component (such as FFA and tocopherols), leaving the TG and carotenes in the raffinate. Upon leaving the extractor, the CO_2-high volatile

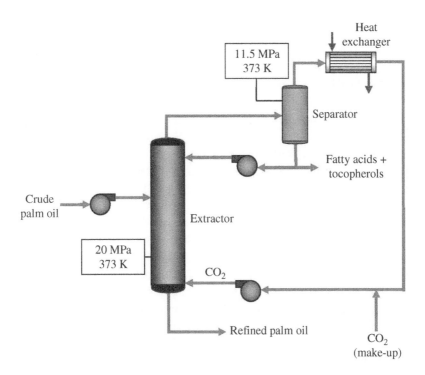

Figure 7.8 Flowsheet of the countercurrent SFE process for palm oil refining.

Table 7.6 Composition of the product obtained from a countercurrent extraction with supercritical CO_2.

Feed/product stream	FFA (wt%)	α-Tocopherol (mg/kg)	β-Carotene (mg/kg)
Feed (CPO)	4.5	1000	540
Extract	85.97	9696	0.0015
Raffinate	0.1	538	568

component stream is flashed in a separator where the FFA and tocopherols are concentrated in the bottom with nearly pure CO_2 in the top product. The recycled CO_2 is mixed with the make-up stream and recompressed prior to entering the extractor. Table 7.6 shows the composition of the product obtained for the countercurrent SFE process simulation.

7.5 Conclusions

Experimental work on deacidification of fats and oils with supercritical CO_2 has been reported by several authors. However, the problems of process optimization and design with both reliable thermodynamic models and mathematical programming techniques have not been addressed. In this study, a conceptual plant model for the refining process of CPO utilizing CO_2 as a supercritical solvent was implemented in the commercial process simulation software, Aspen Plus® 10.2.1 to gain a better understanding on the selection of process conditions for these nonconventional separation processes.

The optimal process scheme and operating conditions for the refining of CPO with supercritical CO_2 were determined through the integration of rigorous process unit models, mathematical programming techniques and predictions of phase equilibrium with the RKA equation of state. In principle, supercritical fluid such as CO_2 is a suitable solvent for the deacidification of crude palm oil. The fatty acid content of crude palm oil can be easily reduced from 4.5 to 0.1 wt%, to an extent that is suitable for edible purposes. The bulk of FFA and vitamin E (as tocopherols and tocotrienols) was extracted in supercritical CO_2, whereas carotenes generally remain in the oil phase.

Part II: ANN, GA Statistical Optimization

7.6 Introduction

Optimization is the act of obtaining the best possible result or the effort for achieving the optimal solution under a given set of circumstances. In designing, development, processing, operation, and maintenance of engineering systems common goals are either to minimize the cost or maximize the desired profits (product quality and operation yield). Such operations should be made efficient by applying the relevant optimization methods and taking the appropriate technological and managerial decisions among all possible alternatives.

Considering high pressures in supercritical extraction processes and sometimes high temperatures, figuring out the right conditions for this process to operate is very important. In case of optimization, sometimes only the reactor is considered in optimization, and in some cases, the whole SFE process has been considered during the optimization process. The objective function for optimization also has been different. Extraction yield, extraction revenue, whole process profit and energy consumption have been the common optimization objective functions. Process constraints have been mostly minimum temperature and pressure to ensure the process is performed in supercritical or subcritical range.

The main process conditions which must be determined to maximize the amount of extract are the following: particle diameter, solvent flow rate, temperature, and pressure. The optimal particle size should be as small as possible but too small particles will stick together and cause solvent channeling in the extractor. The optimal solvent flow rate should be as large as possible. The channeling problem also must be considered. Regarding the optimum temperature and pressure, the problem is more complicated because both the distribution coefficient and the amount of extractable oil change consequently with the system conditions.

7.7 Traditional Optimization

In order to perform optimization, correlations or programs are necessary for prediction of the distribution coefficient and initial mass fraction of oil (solute) as a function of temperature and pressure. Gradient descent algorithms are used to minimize the objective function in this type of optimization. Special care should be taken to see if the obtained optimum point is the local optimum point or the global optimum point. The traditional optimizations can be executed in different software. MATLAB is among the powerful programming languages which have the capability of traditional and nontraditional optimization techniques.

The programs of "griddata," for correlating the initial mass fraction data as a function of temperature and pressure, and the program of "griddata3," for correlating the distribution coefficient data as a function of temperature, pressure, and fluid mass flow rate, in MATLAB software (MATLAB 2004) have been used for this purpose "griddata" and "griddata3" fits a surface and a hyper surface to the data in the (usually) nonuniformly spaced vectors, respectively. These programs use the specified interpolation method for fitting. These methods are "linear," "cubic," "nearest," and "v4" for "griddata" and "linear" and "nearest" for "griddata3." The method defines the type of surface fit to the data. The "cubic" and "v4" methods produce smooth surfaces, while "linear" and "nearest" have discontinuities in the first and zero's derivatives, respectively. Also, the "v4" method uses the method documented in (Barber et al. 1996), but the other methods are based on a Delaunay triangulation of the data (Su et al. 2016). In a case study of us, the "linear" method has been used for both "griddata" and "griddata3" (Rahimi et al. 2011).

Figure 7.9 reveals the influence of temperature and pressure on the amount of extract. From Figure 7.9, it is clear that the extraction yield is maximized at

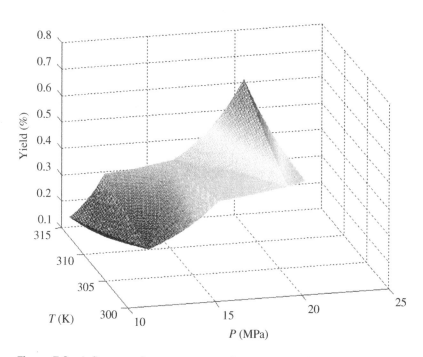

Figure 7.9 Influence of temperature and pressure on the extraction yield at $L = 0.1655$ m, $D_{ext} = 0.0396$ m, $Q = 6.67 \times 10^{-5}$ kg/s at 20 minutes (*Source:* Rahimi et al. (2011). © Elsevier).

optimum operational conditions. The optimal conditions obtained are temperature of 313.15 K and pressure of 20 MPa.

The influence of mass flow rate on the optimum value of temperature and pressure is shown in Figures 7.10 and 7.11, respectively. Figure 7.10 indicates that the optimum temperature is a strong function of mass flow rate, and it increases linearly with mass flow rate. The optimum temperature increased from 303.15 to 313.15 K when mass flow rate increased from 3.33×10^{-5} to 6.67×10^{-5} kg/s. Unlike optimum temperature, the optimum pressure is a weak function of mass flow rate.

In this case study, supercritical fluid extraction of oil from chamomile has been investigated. The experiment of SFE has been performed at 315.15 K, 11.55×10^{-5} kg/s and pressures of 25 MPa. In the next part of the paper, a mathematical modeling has been employed using conservation of mass.

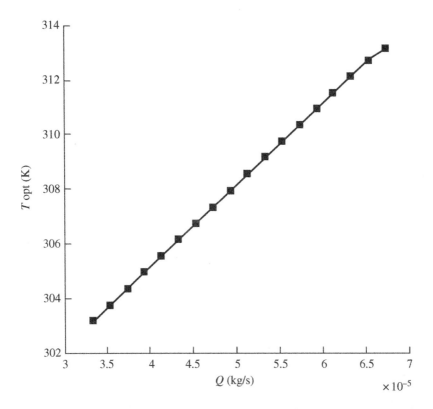

Figure 7.10 Influence of mass flow rate on the optimum temperature at L = 0.1655 m, D_{ext} = 0.0396 m and 20 minutes (*Source: Rahimi et al. (2011).* © *Elsevier*).

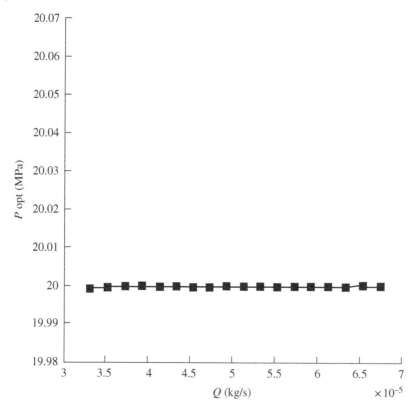

Figure 7.11 Influence of mass flow rate on the optimum pressure at $L = 0.1655$ m, $D_{ext} = 0.0396$ m, and 20 minutes (*Source: Rahimi et al. (2011). © Elsevier*).

Two partial differential equations were obtained for solute concentration: one in fluid and the other for solid phase. The lumped-system assumption has been applied for the particles. The only adjustable parameter in the model is the distribution coefficient of the extract between the fluid and solid phases, which was obtained by optimization.

7.8 Nimbin Extraction Process Optimization

In another case study, the whole process of SFE was selected to be optimized (Zahedi et al. 2011) see Fig. 7.12. Based on our experience and the practical allowance of changing process parameters, the following was selected as

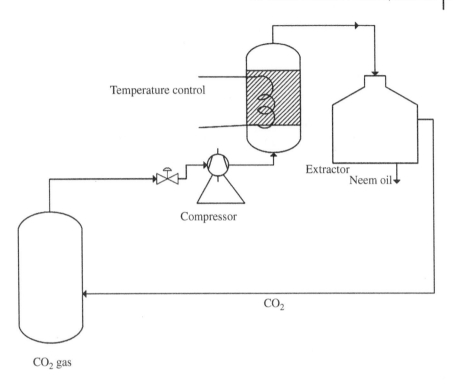

Figure 7.12 Schematic diagram of supercritical CO_2 extraction system.

the optimization variables: temperature, pressure, carbon dioxide flow rate, and particle size. The objective function was set as maximizing profit which is more realistic and more comprehensive than the maximization of cumulative neem production rate (see Fig 7.12). The profit consists of five parts:

$$\text{profit} = C1 - C2 - C3 - C4 - C5 \tag{7.1}$$

$$C1 = \text{yield} * p_1; \tag{7.2}$$

$$C2 = P_{wc} * \text{time} * p_2; \tag{7.3}$$

$$C3 = P_{wcoo} * \text{time} * p_2; \tag{7.4}$$

$$C4 = p_3 * \text{Qm} * \text{time}; \tag{7.5}$$

$$C5 = \text{Cost4} * \text{wt}; \tag{7.6}$$

where yield is mass of extracted nimbin. This quantity can be calculated from F. p_1 is nimbin price. P_{wc} is real compressor work to provide supercritical carbon dioxide and circulate it. The quantity was calculated using Eq. 7.7 as follows:

$$P_{wc} = \frac{1}{440} \frac{\gamma}{\gamma-1} \sqrt{z_1 z_2} N_s P_1 V_1 \left[\left[\frac{P_2}{P_1}\right]^{\frac{\gamma-1}{\gamma}} - 1\right] \quad (7.7)$$

where p_2 is the price of electricity, and P_{wco} is the work necessary to cool CO_2. This process is necessary to remove extracted nimbin from carbon dioxide.

$$P_{wco} = \frac{1}{440} \frac{\gamma}{\gamma-1} \sqrt{z_1 z_2} P_1 V_1 \left[\left[\frac{P_2}{P_1}\right]^{\frac{\gamma-1}{\gamma}} - 1\right] \quad (7.8)$$

p_3 is the price of liquid CO_2. p_1 is neem seed price.

Higher pressures provide economical operation. Effect of carbon dioxide flow rate on process profit is interesting. At lower extraction times, higher CO_2 flow rates should be used. As time goes on, after about 250 minutes, the lower carbon dioxide flow should be used to gain better profits. For higher profits, extraction should be carried out at lower temperatures (Table 7.7).

For the optimization purpose temperature, pressure, CO_2 flow rate, and particle size have been selected as optimization variables. Process limitations have been considered as constraints in optimization routines. GA and GS methods both were used to optimize the process. GS methods provide more accurate results than GA, but GA was faster than GS method.

Sensitivity analysis is used to ascertain how a given model output depends upon the input parameters. The Nonlinear Programming (NLP) problem is performed:

$$\text{Maximize}: \text{profit}(x_i, p_j) \quad (7.9)$$

Table 7.7 Operating variables and constraints.

Independent variables	Lower-upper bounded variable
Pressure (bar)	100–260
Temperature (K)	305–340
CO_2 flow rate (cm^3/min)	0.24–1.24
Particle size (cm)	0.0575–0.1850
Amount of neem used (g)	1–2.5
Extraction time (min)	≥ 0

Subject to

$$h_k(x_i, p_j) = 0 \tag{7.10}$$

$$g_l(x_i, p_j) \geq 0 \tag{7.11}$$

where x_i is decision variables, and p_j is model parameters. The effect of parameter change on the objective function and decision variable can be expressed as follows:

$$S_{p_j} = \left\{ \frac{\dfrac{\partial \operatorname{profit}(x_i, p_j)}{\partial p_j}}{\dfrac{\partial x_j}{\partial p_j}} \right\} \tag{7.12}$$

where S_{pj} is the target of sensitivity of a single item.

Six parameters (p_j, $j = 1, ..., 6$) that affect the profit would be considered as the following:

Price of nimbin, price of CO_2, price of neem, price of electricity, value of coefficient (C) in mass transfer correlation ($Sh = CRe^{0.5} Sc^{0.33}$), value of coefficient "C_1" in mass transfer correlation ($Sh = 0.084\, Re^{\,C_1} Sc^{0.33}$).

Results of sensitivity analysis show that the profit is only sensitive to selling price of nimbin, because the price of nimbin is very high compared with parameters of other prices and values. It was also found that a change in the parameters does not affect the location of the optimum decision or operating variables (pressure, temperature, CO_2 flow rate, size of particle, and amount of neem used, but they have a different optimum operating time when change in the parameters occur (Figure 7.13 and Table 7.8).

7.9 Genetic Algorithm for Mass Transfer Correlation Development

Genetic algorithms can be employed also for developing mass transfer correlation (Zahedi et al. 2010). In a study to model and optimize clove oil extraction based on previous studies, it was found that the mass transfer correlation has the following general form (Mongkholkhajornsilp et al. 2005; Tan et al. 1988; Wakao and Funazkri 1978).

$$Sh = C_1\, Re^{\,C_2} Sc^{\,C_3} \tag{7.13}$$

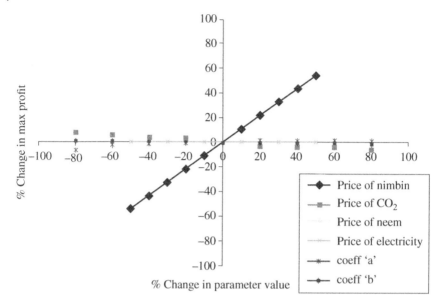

Figure 7.13 Sensitivity analysis of the profit.

Table 7.8 Optimum operating condition from sensitivity analysis.

Variable	Optimum value
Pressure (bar)	196
Temperature (K)	307
CO_2 flow rate (cm³/min)	0.261
Particle size (cm)	0.0575
Amount of neem used (g)	2.5

The value of constants should be determined so that the model fits well the experimental data. In other words, these values should be determined so that the sum of the squares of the difference between the experimental extraction yield and the predicted extraction yield is minimized. This criterion is described in Eq. 7.14:

$$\text{OBJ} = \sum \left(F_{\text{model}} - F_{\text{exp}} \right)^2 \qquad (7.14)$$

The procedure consists of the following steps:

1) Definition of the fitness function. In this chapter, the fitness function is OBJ which has been defined in Eq. 7.14.
2) Characterization of GA parameters (population size: 500, generations: 400, selection options: stochastic uniform, crossover options: scattered, crossover fraction: 0.8 and mutation options).
3) Gaussian.
4) Production of initial generation in a random way.
5) Fitness evaluation.
6) Reproduction a new generation using the GA selection operator.
7) Obtaining crossover pairs of members in the new generation.
8) Performing mutation in the new generation.
9) Steps 4–7 should be repeated until the number of generations reaches the prescribed value.

By implementing GA technique, the following correlation was obtained to correlated mass transfer coefficient with system variables:

$$Sh = 0.0306 \, Re^{0.75} Sc^{0.33} \tag{7.15}$$

7.10 Optimizing Chamomile Extraction

In another study (Rahimi et al. 2011), the chamomile extraction process was investigated and optimized. Based on the operational conditions, the main process conditions which must be determined to maximize the extract are particle diameter, solvent flow rate, temperature, and pressure. The optimal particle size should be as small as possible, but it should not lead to a stick of small particles with consequent solvent channeling in the extractor. The optimal solvent flow rate should be as large as possible, but keeping a minimum solvent residence time so that the solvent can interact with the raw material, the channeling problem also must be considered.

In the case of the optimum temperature and pressure, the problem is more complicated because both the distribution coefficient and the amount of extractable material change with the system operating conditions. Before the optimization, correlations or programs are necessary for predicting the distribution coefficient and initial mass fraction of chamomile extractable material as a function of temperature and pressure. The program, "griddata," for correlating the initial mass fraction data as a function of temperature and pressure, and the program, "griddata3," for correlating the distribution coefficient data as a function of temperature, pressure, and fluid mass flowrate in MATLAB software have been used for this purpose, "griddata" and "griddata3" fit a surface and a hyper surface, respectively, to the data in the (usually) nonuniformly spaced vectors. These programs use the specified interpolation method for fitting. These methods are "linear," "cubic," "nearest" and "v4" for "griddata" and "linear" and "nearest" for "griddata3." The method defines the type of surface fit to the data. The "cubic" and "v4" methods produce smooth surfaces, while "linear" and "nearest" have discontinuities in the first and zero's derivatives, respectively. The optimal conditions obtained by the genetic algorithm are temperature of 313 K and pressure of 20 MPa.

The effects of mass flow rate on the optimum value of temperature and pressure are shown in Figures 7.14 and 7.15, respectively. The effect of mass flow rate on the optimum value of temperature and pressure are shown in Figure 7.16. The optimum temperature increased from 303.15 to 313.15 K when mass flow rate increased from 3.33×10^{-5} to 6.67×10^{-5} kg/s. Unlike optimum temperature, the optimum pressure is a weak function of mass flow rate.

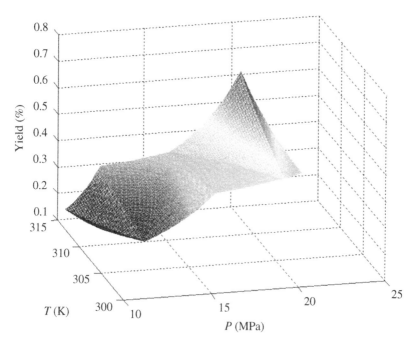

Figure 7.14 Effect of temperature and pressure on the extraction yield at $L = 0.1655$ m, $D_{ext} = 0.0396$ m, $Q = 6.67 \times 10^{-5}$ kg/s at 20 minutes.

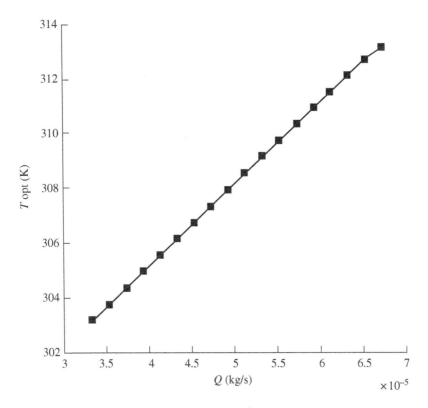

Figure 7.15 Effect of mass flow rate on the optimum temperature at L = 0.1655 m, D_{ext} = 0.0396 m, and 20 minutes.

7.11 Statistical and ANN Optimization

In optimization practice, sometimes the question is should we use ANN for optimization or statistical techniques? In the following parts, different studies and comparison between these two techniques have been provided. In the first study (Zahedi and Azarpour 2011) to make sure that the selected data for modeling present normal operating ranges, the unsatisfactory data were excluded from the data bank.

In order to make ANN model, the back-propagation learning with one hidden-layer network was used. Inputs and outputs are normalized between the values −1 and 1. Logistic Sigmoid and purelin transfer functions have been used in constructing ANNs. ANN has been trained with 70% of the data set and 30% of the data have been applied for testing the predictions of ANN.

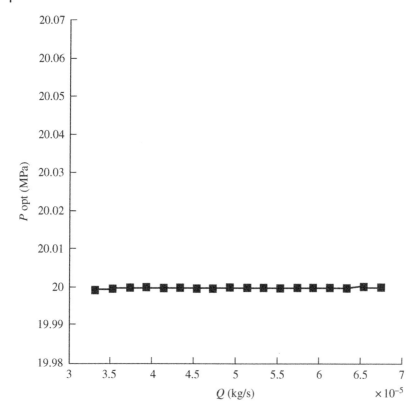

Figure 7.16 Influence of mass flow rate on the optimum pressure at $L = 0.1655$ m, $D_{ext} = 0.0396$ m, and 20 minutes.

In order to build a Response Surface Model (RSM), the response variable is the oil extraction yield, and the process variables are temperature, pressure, and extraction time. In this case, a full quadratic model was selected. Then, based on these results, the mathematical model which represents the extraction process was obtained as follows:

$$Y = -71.453 + 2.5348\,X_1 - 0.0218\,X_1^2 + 1.3856\,X_2 - 0.0266\,X_2^2 \\ + 4.276\,X_3 - 0.1635\,X_3^2 + 0.0021\,X_1X_2 - 0.0385\,X_1X_3 - 0.0323\,X_2X_3 \tag{7.16}$$

Figure 7.17 shows 3D results and optimum point.

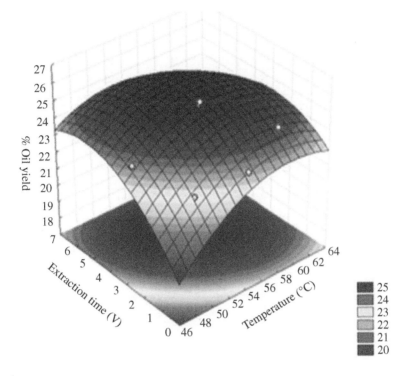

Figure 7.17 Three-dimensional plot of oil yield as a function of temperature and extraction time.

To summarize the results, the optimum extraction process parameters within the experimental ranges are; temperature of 55.9 °C, pressure of 25.8 MPa, and extraction time of 3.95 hours. Under these conditions, the oil extraction yield will be 25.76%.

In order to compare optimization results of ANN and RSM, relative error has been used to evaluate the results and analogy as well. The percent relative error is defined as follows:

$$\text{Percent relative error} = \frac{\text{value} - \text{approximate}}{\text{value}} \times 100 \qquad (7.17)$$

where "value" is the experimental value used to construct the model and "approximate" is the output of the neural networks at the same conditions. Regardless of the model, this error can have either positive or negative values. However, the better result is the convergence of relative error parameters to zero. Errors of measurements using this criterion have been shown in Table 7.9.

Table 7.9 Comparison of the ANN prediction with RSM for training data.

Temperature (°C)	Pressure (MPa)	Extraction time (h)	Oil yield (exp.)	Predicted value (ANN)	Predicted value (RSM)	Relative error (%) (ANN)	Relative error (%) (RSM)
60	30	5	24.7	24.700	24.487	−0.000190	0.86032
60	30	1	24.39	24.389	24.423	0.000323	−0.13735
60	20	1	23.14	23.140	22.921	−0.000020	0.94425
50	30	1	22.68	22.679	22.8105	0.000002	−0.57539
50	20	5	24.44	24.440	24.4145	−0.000184	0.10433
50	20	1	21.3	21.299	21.5185	0.000201	−1.02582
63	25	3	24.27	24.269	24.7177	0.000034	−1.84466
47	25	3	23.78	23.780	23.5369	0.000068	1.02228
55	25	6.36	24.75	24.749	24.9199	0.000081	−0.68646
55	25	0.36	23.4	23.400	23.4062	−0.000820	−0.02649
55	25	3	25.6	25.599	25.5225	0.000450	0.30273

It is obvious that ANN has a superior overlap with the laboratory experimental data compared to RSM. Mean Square Error values for ANN model and RSM method were calculated 0.0009 and 0.0421, respectively. Therefore, the results derived from the ANN model were significant and the error was incredibly low.

It can be observed that the optimum extraction process parameters within the experimental ranges are temperature of 56.50 °C, pressure of 23.30 MPa, and extraction time of 3.72 hours. Under these conditions, the oil extraction yield was 26.55% while the yield obtained from RSM model was 25.76%. As ANN is more accurate than RSM, it can be concluded that the optimum values from ANN model optimization are more reliable.

The same investigation was performed in another study (Pouralinazar et al. 2012). The research is for subcritical fluid extraction of *Orthosiphon stamineus*. In this study Box–Behnken experimental design method and ANN were developed to optimize the extraction process. Table 7.10 depicts the range of variables and designed experiments.

In this study, temperature in the quadratic form had the most dominant effect on the oil yield. The model equation representing the response (z) was expressed as function of temperature (x_1), extraction time (x_2), extraction cycle (x_3) for coded unit as the following:

$$Z = -217.085 + 3.816\,x_1 + 5.2703\,x_2 + 17.26\,x_3 - 0.0179\,x_1^2$$
$$- 0.193\,x_2^2 - 3.717\,x_3^2 - 0.01625\,x_1 x_3 - 0.01539\,x_1 x_2 + 0.118\,x_2 x_3$$
$$(7.18)$$

Table 7.10 Experimental range and levels of independent variables.

	Range and level		
Variables	Low level (−1) ΔX_i^a	Center level (0)	High level (+1)
Temperature (°C)	80	100	120
		20	
Extraction time (min)	5	10	15
		5	
Extraction cycle	1	2	3
		1	

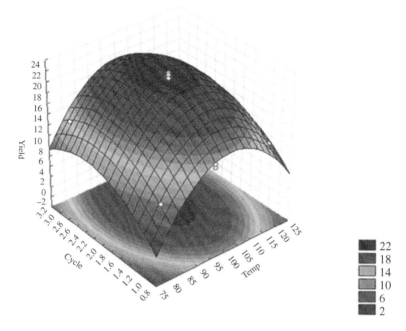

Figure 7.18 Effect of temperature and extraction cycle on the oil yield (contour plot).

Figure 7.18 shows a typical counter plot of the following model.

7.12 Conclusion

In sub- and supercritical optimization, the important factor is setting up proper objective function and process constraints. If the accurate model of the system is not in hand, artificial intelligence models provide better modeling and optimization results. When the first principle model of the system is available, usually GA techniques provide faster results. Finally, if traditional optimization techniques are applied, the optimum point should be examined to assess the global optimality of the solution.

Appendix A

Calculation of the Composition for Palm Oil TG (Lim et al. 2003)

The composition of palm oil reported by Tan and Oh (1981) is given in Table 3.2. The major fatty acids of palm oil TG are palmitic acid and oleic acid, with compositions of 43.5 and 39.8 wt%, respectively. Normalized, the composition of both fatty acids is as follows:

$$\text{Composition of tripalmitin} = \frac{x_{16:0}}{\sum x_{16:0} + x_{18:0}}$$

$$= \frac{43.5}{(43.5 + 39.8)}$$

$$= 52.22 \text{ wt\%}$$

$$\text{Composition of triolein} = \frac{x_{18:0}}{\sum x_{16:0} + x_{18:0}}$$

$$= \frac{39.8}{(43.5 + 39.8)}$$

$$= 47.78 \text{ wt\%}$$

Modeling, Simulation, and Optimization of Supercritical and Subcritical Fluid Extraction Processes,
First Edition. Zainuddin A. Manan, Gholamreza Zahedi, and Ana Najwa Mustapa.
© 2022 by the American Institute of Chemical Engineers, Inc.
Published 2022 by John Wiley & Sons, Inc.

Appendix B

Calculation of Distribution Coefficient and Separation Factor (Lim et al. 2003)

The simulated liquid-phase and supercritical fluid-phase compositions were used as the basis for calculating the distribution coefficient and separation factor of palm oil components. Figure B.1 shows the data generated from a flash calculation (at 30 MPa and 343 K) using the Aspen Plus® 10.2.1 process simulator. Transformation of the multicomponent data on a CO_2-free basis is given in Table B.1.

Distribution coefficients of palm oil components were calculated by Eq. 4.27:

$$K_{FFA} = 0.2677/0.0437$$
$$= 6.1321$$

$$K_{TGc} = (0.4644 + 0.2653)/(0.4653 + 0.4895)$$
$$= 0.7643$$

$$K_{Tocopherol} = 0.0022/0.0010$$
$$= 2.2610$$

$$K_{Carotene} = 0.0003/0.0005$$
$$= 0.5201$$

Separation factor between palm oil components was calculated according to Eq. 4.28:

$$\alpha_{FFA/TG} = K_{FFA}/K_{TG}$$
$$= 6.1321/0.7643$$
$$= 8.0232$$

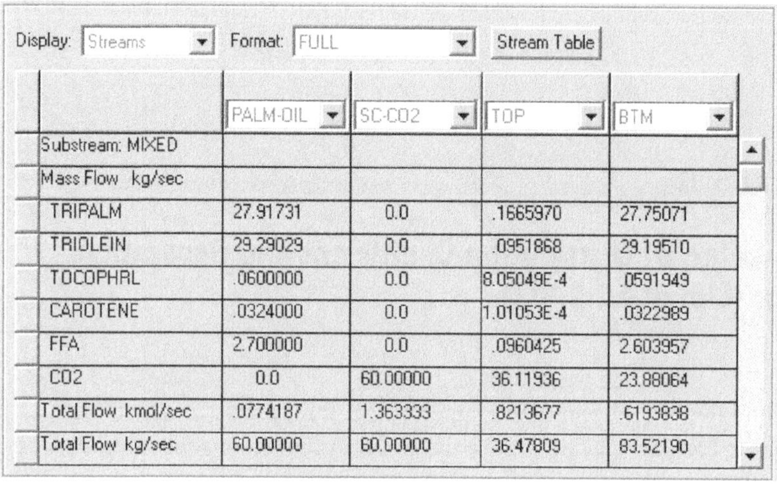

Display: Streams Format: FULL Stream Table

	PALM-OIL	SC-CO2	TOP	BTM
Substream: MIXED				
Mass Flow kg/sec				
TRIPALM	27.91731	0.0	.1665970	27.75071
TRIOLEIN	29.29029	0.0	.0951868	29.19510
TOCOPHRL	.0600000	0.0	8.05049E-4	.0591949
CAROTENE	.0324000	0.0	1.01053E-4	.0322989
FFA	2.700000	0.0	.0960425	2.603957
CO2	0.0	60.00000	36.11936	23.88064
Total Flow kmol/sec	.0774187	1.363333	.8213677	.6193838
Total Flow kg/sec	60.00000	60.00000	36.47809	83.52190

Figure B.1 Calculated liquid-phase and supercritical fluid-phase compositions at 30 MPa and 343 K using the Aspen Plus 10.2.1 process simulator.

Table B.1 Multicomponent data transformed to a CO_2-free basis.

	Fluid phase		Liquid phase	
Component	**(kg/s)**	**(wt%)**	**(kg/s)**	**(wt%)**
Tripalmitin	0.1666	0.4644	27.7507	0.4653
Triolein	0.0952	0.2653	29.1951	0.4895
Tocopherol	0.0008	0.0022	0.0592	0.0010
Carotene	0.0001	0.0003	0.0323	0.0005
FFA	0.0960	0.2677	2.6040	0.0437
Total (CO_2-free)	0.3587	1.0000	59.6413	1.0000

Appendix C

Calculation of Palm Oil Solubility in Supercritical CO_2 (Lim et al. 2003)

Assumed extraction − stage efficiency = 50%

Solvent-to-feed (S/F) ratio used in the study of Ooi et al. (1996) = 40

Crude palm oil feed rate = 60 g/s

Supercritical CO_2 required (under continuous operation) = $0.5 \times 40 \times 60$
$$= 1.2 \text{ g/s}$$

Flash calculation results conducted at the above conditions (24 MPa, 50 °C) are given in Figure C.1.

Palm oil solubility in supercritical CO_2 = $(1.312415) - (1.297979)$ kg/s
$$= 14.4363 \text{ g/s}$$

Palm oil solubility (under continuous operation) = $\dfrac{14.4363 \text{ g oil/s}}{(40 \times 60) \text{ g } CO_2/s}$
$$\times 100\% = 0.6015\%$$

Modeling, Simulation, and Optimization of Supercritical and Subcritical Fluid Extraction Processes,
First Edition. Zainuddin A. Manan, Gholamreza Zahedi, and Ana Najwa Mustapa.
© 2022 by the American Institute of Chemical Engineers, Inc.
Published 2022 by John Wiley & Sons, Inc.

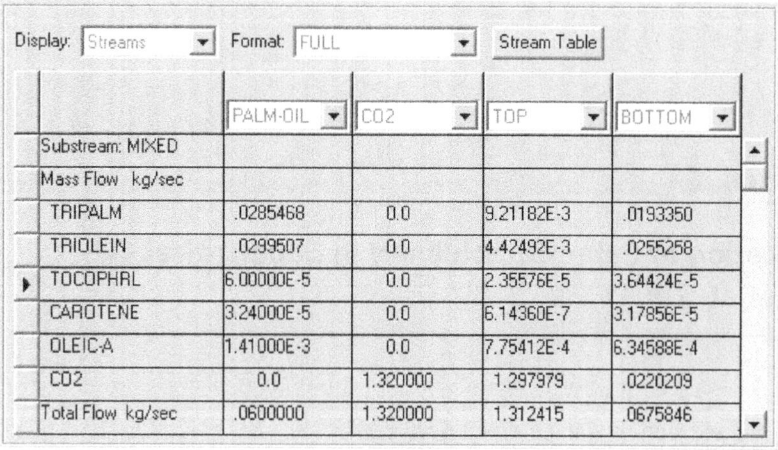

Display: Streams Format: FULL Stream Table

	PALM-OIL	CO2	TOP	BOTTOM
Substream: MIXED				
Mass Flow kg/sec				
TRIPALM	.0285468	0.0	9.21182E-3	.0193350
TRIOLEIN	.0299507	0.0	4.42492E-3	.0255258
TOCOPHRL	6.00000E-5	0.0	2.35576E-5	3.64424E-5
CAROTENE	3.24000E-5	0.0	6.14360E-7	3.17856E-5
OLEIC-A	1.41000E-3	0.0	7.75412E-4	6.34588E-4
CO2	0.0	1.320000	1.297979	.0220209
Total Flow kg/sec	.0600000	1.320000	1.312415	.0675846

Figure C.1 Flash calculation results at 24 MPa and 50 °C.

References

Abbott, A.P. and Eardley, C.A. (1998). Solvent properties of liquid and supercritical 1, 1, 1, 2-tetrafluoroethane. *The Journal of Physical Chemistry B 102* (43): 8574–8578.

Abbott, A.P., Eardley, C.A., and Scheirer, J.E. (1999). Solvent properties of supercritical CO_2/HFC134a mixtures. *The Journal of Physical Chemistry B 103* (41): 8790–8793.

Ajchariyapagorn, A., Kumhom, T., Pongamphai, S. et al. (2009). Predicting the extraction yield of nimbin from neem seeds in supercritical CO_2 using group contribution methods, equations of state and a shrinking core extraction model. *The Journal of Supercritical Fluids 51* (1): 36–42.

Al-Otoom, A., Al-Asheh, S., Allawzi, M. et al. (2014). Extraction of oil from uncrushed olives using supercritical fluid extraction method. *The Journal of Supercritical Fluids 95*: 512–518.

Araújo, M.E., Machado, N.T., and Meireles, M.A.A. (2001). Modelling the phase equilibrium of soybean oil deodorizer distillates + supercritical carbon dioxide using the Peng-Robinson EOS. *Industrial & Engineering Chemistry Research 40* (4): 1239–1243.

Ashour, I. (1989). Supercritical fluid extraction of fatty acids and their derivatives: Solubility measurements, thermodynamic modelling and process simulation. Doctoral thesis. Lund University.

Ashraf-Khorassani, M., Combs, M.T., Taylor, L.T. et al. (1997). Solubility study of sulfamethazine and sulfadimethoxine in supercritical carbon dioxide, fluoroform, and subcritical freon 134A. *Journal of Chemical & Engineering Data 42* (3): 636–640.

Aspen Technology, Inc (2000). *ASPEN PLUS® User Guide Vol. 1, Release 10.2.1*. Cambridge, MA: Aspen Technology.

Azarpour, A., Alwi, S.R.W., Zahedi, G. et al. (2015). Catalytic activity evaluation of industrial Pd/C catalyst via gray-box dynamic modeling and simulation of hydropurification reactor. *Applied Catalysis A: General 489*: 262–271.

Azizi, C.Y. (2007). The identification of extraction and separation of vitamin E and Djenkolic acid from Pithecellobium Jiringan (Jack) prain seeds using supercritical carbon dioxide. PhD thesis. Universiti Sains Malaysia.

Baharuddin, N.S., Abdullah, H., and Wahab, W.N.A.W.A. (2015). Anti-Candida activity of Quercus infectoria gall extracts against Candida species. *Journal of Pharmacy & Bioallied Sciences 7* (1): 15.

Balachandran, S., Kentish, S., and Mawson, R. (2006). The effects of both preparation method and season on the supercritical extraction of ginger. *Separation and Purification Technology 48* (2): 94–105.

Bamberger, T., Erickson, J.C., Cooney, C.L., and Kumar, S.K. (1988). Measurement and model prediction of solubilities of pure fatty acids, pure triglycerides, and mixtures of triglycerides in supercritical carbon dioxide. *Journal of Chemical and Engineering Data 33* (3): 327–333.

Barber, C.B., Dobkin, D.P., and Huhdanpaa, H.T. (1996). The quickhull algorithm for convex hulls. *ACM Transactions on Mathematical Software 22* (4): 469–483.

Barker, P., Cary, R., Dobson, S., and Organization W H (1999). *1, 1, 1, 2-Tetrafluoroethane*. World Health Organization.

Bashipour, F. and Ghoreishi, S.M. (2014). Response surface optimization of supercritical CO_2 extraction of α-tocopherol from gel and skin of Aloe vera and almond leaves. *The Journal of Supercritical Fluids 95*: 348–354.

Běhounek, L. and Cintula, P. (2006). Fuzzy logics as the logics of chains. *Fuzzy Sets and Systems 157* (5): 604–610.

Berchmans, H.J. and Hirata, S. (2008). Biodiesel production from crude Jatropha curcas L. seed oil with a high content of free fatty acids. *Bioresource Technology 99* (6): 1716–1721.

Bezerra, M.A., Santelli, R.E., Oliveira, E.P. et al. (2008). Response surface methodology (RSM) as a tool for optimization in analytical chemistry. *Talanta 76* (5): 965–977.

Bharath, R. (1993). Measurement and prediction of phase equilibria of supercritical CO_2-natural fat and oil component systems. Doctoral thesis. Tohoku University, Sendai, Japan.

Bharath, R., Inomata, H., Adschiri, T., and Arai, K. (1992). Phase equilibrium study for the separation and fractionation of fatty oil components using supercritical carbon dioxide. *Fluid Phase Equilibria 81*: 307–320.

Bharath, R., Yamane, S., Inomata, H. et al. (1993). Phase equilibria of supercritical CO_2-fatty oil component binary systems. *Fluid Phase Equilibria 83*: 183–192.

Bhattacharjee, P., Singhal, R.S., and Tiwari, S.R. (2007). Supercritical carbon dioxide extraction of cottonseed oil. *Journal of Food Engineering 79* (3): 892–898.

Biegler, L.T. and Cuthrell, J.E. (1985). Improved infeasible path optimisation for sequential modular simulators – II: the optimisation algorithm. *Computers and Chemical Engineering 9* (3): 257–267.

R. Byron Bird, Warren E. Stewart, and Edwin N. Lightfoot. (1924). *Transport Phenomena*, 2e. Wiley.

Birtigh, A., Johannsen, M., Brunner, G., and Nair, N. (1995). Supercritical-fluid extraction of oil-palm components. *The Journal of Supercritical Fluids 8* (1): 46–50.

Bishop, C.M. (1995). *Neural Networks for Pattern Recognition*. Oxford University Press.

Bisunadan, M.M. (1993). Extraction of oil from oil palm fruits using supercritical carbon dioxide. MSc thesis. Universiti Sains Malaysia, Malaysia.

Bitner, E., Friedrich, J., and Mounts, T. (1986). Laboratory continuous deodorizer for vegetable oils. *Journal of the American Oil Chemists Society 63* (3): 338–340.

Blackwell, J.A., Chen, D.T., Alband, T.D., and Perman, C.A. (1996). Supercritical fluid extraction involving hydrofluoroalkanes. Patent 5,481,058, filed 7 January 1994.

Bloemen, F. (1966). Column for steam refining and deodorization of fats and for the simple or fractional distillation of fatty acids. *Revue Francaise des Corps Gras 13*: 247–254.

Bocevska, M. and Sovová, H. (2007). Supercritical CO_2 extraction of essential oil from yarrow. *The Journal of Supercritical Fluids 40* (3): 360–367.

Bollas, G., Papadokonstantakis, S., Michalopoulos, J. et al. (2004). A computer-aided tool for the simulation and optimization of the combined HDS-FCC processes. *Chemical Engineering Research and Design 82* (7): 881–894.

Bondioli, P., Mariani, C., Lanzani, A. et al. (1992). Lampante olive oil refining with supercritical carbon dioxide. *Journal of the American Oil Chemists Society 69* (5): 477–480.

Box, G.E.P., Hunter, W.G., and Hunter, J.S. (1978). *Statistics for Experimenters: An Introduction to Design, Data Analysis and Model Building*. New York: Wiley.

Brachet, A., Christen, P., Gauvrit, J.-Y. et al. (2000). Experimental design in supercritical fluid extraction of cocaine from coca leaves. *Journal of Biochemical and Biophysical Methods 43* (1–3): 353–366.

Bravi, M., Spinoglio, F., Verdone, N. et al. (2007). Improving the extraction of α-tocopherol-enriched oil from grape seeds by supercritical CO_2. Optimisation of the extraction conditions. *Journal of Food Engineering 78* (2): 488–493.

Brunetti, L., Daghetta, A., Fedell, E. et al. (1989). Deacidification of olive oils by supercritical carbon dioxide. *Journal of the American Oil Chemists Society 66* (2): 209–217.

Brunner, G. (1978). Phase equilibria in the presence of compressed gases and their importance in the separation of low-volatile materials. Post-doctoral thesis. Universität Erlangen-Nürnberg.

Brunner, G. (1994). *Gas Extraction: An Introduction to Fundamentals of Supercritical Fluids and the Application to Separation Processes*. New York: Steinkopff Darmstadt Springer.

Brunner, G. (1998). Industrial process development countercurrent multistage gas extraction (SFE) processes. *Journal of Supercritical Fluids 13* (1–3): 283–301.

Brunner, G. (2000). Fractionation of fats with supercritical carbon dioxide. *European Journal of Lipid Science and Technology 102* (3): 240–245.

Brunner, G. (2005). Supercritical fluids: technology and application to food processing. *Journal of Food Engineering 67*: 21–33.

Brunner, G. and Peter, S. (1982). State of art of extraction with compressed gases (gas extraction). *German Chemical Engineering 5*: 181–195.

Brunner, G., Malchow, T., Stürken, K., and Gottschau, T. (1991). Separation of tocopherols from deodorizer condensates by countercurrent extraction with carbon dioxide. *The Journal of Supercritical Fluids 4* (1): 72–80.

Bruno, T.J. (2006). Experimental approaches for the study and application of supercritical fluids. *Combustion Science and Technology 178* (1–3): 3–46.

Bulley, N., Fattori, M., Meisen, A., and Moyls, L. (1984). Supercritical fluid extraction of vegetable oil seeds. *Journal of the American Oil Chemists Society 61* (8): 1362–1365.

Bulsari, A. (1995). *Neural Networks for Chemical Engineers*. Elsevier Science Inc.

Casas, L., Mantell, C., Rodríguez, M. et al. (2007). Effect of the addition of cosolvent on the supercritical fluid extraction of bioactive compounds from Helianthus annuus L. *The Journal of Supercritical Fluids 41* (1): 43–49.

Castro, M.D.L.d., Valcárcel, M., and Tena, M.T. (1994). *Analytical Supercritical Fluid Extraction*. German: Springer-Verlag.

Catchpole, O.J. and King, M.B. (1994). Measurement and correlation of binary diffusion coefficients in near critical fluids. *Industrial & Engineering Chemistry Research 33* (7): 1828–1837.

Catchpole, O.J. and Proells, K. (2001). Solubility of squalene, oleic acid, soya oil, and deep sea shark liver oil in subcritical R134a from 303 to 353 K. *Industrial and Engineering Chemistry Research 40* (3): 965–972.

Catchpole, O.J., Grey, J.B., and Noermark, K.A. (2000). Fractionation of fish oils using supercritical CO_2 and CO_2 + ethanol mixtures. *Journal of Supercritical Fluids 19* (1): 25–37.

Cavalcanti, R.N., Albuquerque, C.L., and Meireles, M.A.A. (2016). Supercritical CO_2 extraction of cupuassu butter from defatted seed residue: experimental data, mathematical modeling and cost of manufacturing. *Food and Bioproducts Processing* 97: 48–62.

Chakrabarti, P.P. and Jala, R.C.R. (2019). Processing technology of rice bran oil. In: *Rice Bran and Rice Bran Oil* (eds. L.Z. Cheong and X. Xu), 55–95. Elsevier.

Charest, D., Balaban, M., Marshall, M., and Cornell, J. (2001). Astaxanthin extraction from crawfish shells by supercritical CO_2 with ethanol as cosolvent. *Journal of Aquatic Food Product Technology* 10 (3): 81–96.

Chen, C. and Ramaswamy, H. (2002). Modeling and optimization of variable retort temperature (VRT) thermal processing using coupled neural networks and genetic algorithms. *Journal of Food Engineering* 53 (3): 209–220.

Chen, C.-C., Chieh-ming, J.C., and Yang, P.-w. (2000). Vapor–liquid equilibria of carbon dioxide with linoleic acid, α-tocopherol and triolein at elevated pressures. *Fluid Phase Equilibria* 175 (1–2): 107–115.

Cheng, A.T., Calvo, J.R., and Barrado, R.R. (1993). Deodorizing edible oil and/or fat with non-condensible inert gas and recovering a high quality fatty acid distillate. European Patent 513,739, filed 27 December 2012.

Chiu, K.-L., Cheng, Y.-C., Chen, J.-H. et al. (2002). Supercritical fluids extraction of Ginkgo ginkgolides and flavonoids. *The Journal of Supercritical Fluids* 24 (1): 77–87.

Choo, Y.M., Ma, A.N., Yahaya, H. et al. (1996). Separation of crude palm oil components by semipreparative supercritical fluid chromatography. *Journal of the American Oil Chemists' Society* 73 (4): 523–525.

Choo, Y.M., Ma, A.N., Ooi, C.K., and Yusof, B. (1997). Red palm oila new carotene-rich nutritious oil. Paper presented at the National Seminar on Palm Oil Milling, Refining Technology and Quality, Sabah, Malaysia (7–8 August 1997).

Choo, Y.M., Ng, M.H., Ma, A.N. et al. (2005). Application of supercritical fluid chromatography in the quantitative analysis of minor components (carotenes, vitamin E, sterols, and squalene) from palm oil. *Lipids* 40 (4): 429–432.

Chrastil, J. (1982). Solubility of solids and liquids in supercritical gases. *The Journal of Physical Chemistry* 86 (15): 3016–3021.

Chuang, M.-H. and Brunner, G. (2006). Concentration of minor components in crude palm oil. *The Journal of Supercritical Fluids* 37 (2): 151–156.

Cocero, M. and Garcıa, J. (2001). Mathematical model of supercritical extraction applied to oil seed extraction by CO_2+ saturated alcohol-I. Desorption model. *The Journal of Supercritical Fluids* 20 (3): 229–243.

Coelho, J., Pereira, A., Mendes, R., and Palavra, A. (2003). Supercritical carbon dioxide extraction of Foeniculum vulgare volatile oil. *Flavour and Fragrance Journal* 18 (4): 316–319.

Constantinides, A. and Mostoufi, N. (1999). *Numerical Methods for Chemical Engineers with Matlab Applications*. Prentice Hall PTR.

Corr, S. (2002). 1, 1, 1, 2-Tetrafluoroethane; from refrigerant and propellant to solvent. *Journal of Fluorine Chemistry 118* (1–2): 55–67.

Crespo, M.P. and Yusty, M.L. (2005). Comparison of supercritical fluid extraction and Soxhlet extraction for the determination of PCBs in seaweed samples. *Chemosphere 59* (10): 1407–1413.

Curren, M.S.S. and King, J.W. (1997). *Critical Fluids for Oil Extraction– Sampling and Sample Preparation for Food Analysis* (eds. P.J. Wan and P.J. Wakelyn), 869–894. Champaign, IL: AOCS Press.

Cygnarowicz, M.L., Maxwell, R.J., and Seider, W.D. (1990). Equilibrium solubilities of β-carotene in supercritical carbon dioxide. *Fluid Phase Equilibria 59* (1): 57–71.

Da Porto, C., Natolino, A., and Decorti, D. (2014). Extraction of proanthocyanidins from grape marc by supercritical fluid extraction using CO_2 as solvent and ethanol–water mixture as co-solvent. *The Journal of Supercritical Fluids 87*: 59–64.

Da Porto, C., Decorti, D., and Natolino, A. (2016). Microwave pretreatment of Moringa oleifera seed: effect on oil obtained by pilot-scale supercritical carbon dioxide extraction and Soxhlet apparatus. *The Journal of Supercritical Fluids 107*: 38–43.

Davoody, M. (2012). *Neuro Fuzzy and Hybrid Modeling of Supercritical Fluid Extraction of Pimpinella Anisum L. Seed*. Universiti Teknologi Malaysia.

Davoody, M., Zahedi, G., Biglari, M. et al. (2012). Expert and gray box modeling of high pressure liquid carbon dioxide extraction of Pimpinella anisum L. seed. *The Journal of Supercritical Fluids 72*: 213–222.

De Greyt, W.F., Kellens, M.J., and Huyghebaert, A.D. (1999). Effect of physical refining on selected minor components in vegetable oils. *Lipid/Fett 101* (11): 428–432.

De Lucas, A.d., de la Ossa, E.M., Rincón, J. et al. (2002). Supercritical fluid extraction of tocopherol concentrates from olive tree leaves. *The Journal of Supercritical Fluids 22* (3): 221–228.

Dean, J., Liu, B., and Price, R. (1998). Extraction of magnolol from Magnolia officinalis using supercritical fluid extraction and phytosol solvent extraction. *Phytochemical Analysis: An International Journal of Plant Chemical and Biochemical Techniques 9* (5): 248–252.

Dejoye, C., Vian, M.A., Lumia, G. et al. (2011). Combined extraction processes of lipid from Chlorella vulgaris microalgae: microwave prior to supercritical carbon dioxide extraction. *International Journal of Molecular Sciences 12* (12): 9332–9341.

Dohrn, R. and Brunner, G. (1994). An estimation method to calculate Tb, Tc, Pc and ω from the liquid molar volume and the vapor pressure. Paper presented at the Proceedings of the 3rd International Symposium on Supercritical Fluids (17–19 October 1994).

Döker, O., Salgin, U., Yildiz, N. et al. (2010). Extraction of sesame seed oil using supercritical CO_2 and mathematical modeling. *Journal of Food Engineering* 97 (3): 360–366.

Duba, K.S. and Fiori, L. (2015). Supercritical CO_2 extraction of grape seed oil: effect of process parameters on the extraction kinetics. *The Journal of Supercritical Fluids* 98: 33–43.

Dunford, N.T., Teel, J.A., and King, J.W. (2003). A continuous countercurrent supercritical fluid deacidification process for phytosterol ester fortification in rice bran oil. *Food Research International* 36 (2): 175–181.

Duran-Valencia, C., Pointurier, G., Valtz, A. et al. (2002). Vapor– liquid equilibrium (VLE) data for the carbon dioxide (CO_2)+ 1, 1, 1, 2-tetrafluoroethane (R134a) system at temperatures from 252.95 K to 292.95 K and pressures up to 2 MPa. *Journal of Chemical & Engineering Data* 47 (1): 59–61.

Đurđević, S., Milovanović, S., Šavikin, K. et al. (2017). Improvement of supercritical CO_2 and n-hexane extraction of wild growing pomegranate seed oil by microwave pretreatment. *Industrial Crops and Products* 104: 21–27.

Efendy Goon, D., Sheikh Abdul Kadir, S.H., Latip, N.A. et al. (2019). Palm oil in lipid-based formulations and drug delivery systems. *Biomolecules* 9 (2): 64.

Eggers, R. (1996). Supercritical fluid extraction (SFE) of oilseeds/lipids. In: *Supercritical Fluid Technology in Oil and Lipid Chemistry* (eds. J.W. King and G. R. List), 35–60. Champaign, IL: AOCS Press.

Eggers, R., Ambrogi, A., and Schnitzler, J.v. (2000). Special features of SCF solid extraction of natural products: deoiling of wheat gluten and extraction of rose hip oil. *Brazilian Journal of Chemical Engineering* 17 (3): 329–334.

Eglese, R.W. (1990). Simulated annealing: a tool for operational research. *European Journal of Operational Research* 46 (3): 271–281.

Ekart, M.P., Bennett, K.L., Ekart, S.M. et al. (1993). Cosolvent interactions in supercritical fluid solutions. *AICHE Journal* 39 (2): 235–248.

Espinosa, S., Diaz, S., and Brignole, E.A. (2000). Optimal design of supercritical fluid processes. *Computers and Chemical Engineering* 24: 1301–1307.

Espinosa, S., Fornari, T., Bottini, S.B., and Brignole, E.A. (2002). Phase equilibria in mixtures of fatty oils and derivatives with near critical fluids using the GC-EOS model. *The Journal of Supercritical Fluids* 23 (2): 91–102.

Esquível, M.M., Bernardo-Gil, M.G., and King, M.B. (1999). Mathematical models for supercritical extraction of olive husk oil. *The Journal of Supercritical Fluids* 16 (1): 43–58.

Fang, T., Goto, M., Wang, X. et al. (2007). Separation of natural tocopherols from soybean oil byproduct with supercritical carbon dioxide. *The Journal of Supercritical Fluids* 40 (1): 50–58.

Fattori, M., Bulley, N.R., and Meisen, A. (1988). Carbon dioxide extraction of canola seed: oil solubility and effect of seed treatment. *Journal of the American Oil Chemists Society* 65 (6): 968–974.

Fernandes, J., Lisboa, P.F., Simoes, P.C. et al. (2009). Application of CFD in the study of supercritical fluid extraction with structured packing: wet pressure drop calculations. *The Journal of Supercritical Fluids* 50 (1): 61–68.

Fernandez Vecilla, A. (1994). Deodorization of oils using nitrogen gas. *Alimentación, Equipos y Tecnología 13*: 23–26.

Fernández, C.M., Fiori, L., Ramos, M.J. et al. (2015). Supercritical extraction and fractionation of Jatropha curcas L. oil for biodiesel production. *The Journal of Supercritical Fluids* 97: 100–106.

Ferreira, S.L.C., Bruns, R.E., da Silva, E.G.P. et al. (2007). Statistical designs and response surface techniques for the optimization of chromatographic systems. *Journal of Chromatography A 1158* (1–2): 2–14.

Fiori, L., Basso, D., and Costa, P. (2009). Supercritical extraction kinetics of seed oil: a new model bridging the 'broken and intact cells' and the 'shrinking-core'models. *The Journal of Supercritical Fluids* 48 (2): 131–138.

Fogel, D.B. (2000). What is evolutionary computation? *IEEE Spectrum* 37 (2): 26–32.

Formo, M.W., Jungermann, E., Norris, F.A., and Sonntag, N.O.V. (1979). *Bailey's Industrial Oil and Fat Products*, 4the, vol. *1*. New York: Wiley.

Fornari, T., Vicente, G., Vázquez, E. et al. (2012). Isolation of essential oil from different plants and herbs by supercritical fluid extraction. *Journal of Chromatography. A 1250* (34–48): 34–48.

Franca, L.F. and Meireles, M.A.A. (1997). Extraction of oil from pressed palm oil (Elaes guineensis) fibers using supercritical CO_2. *Food Science and Technology* 17 (4): 384–388.

França, L.F.d. and Meireles, M.A.A. (2000). Modeling the extraction of carotene and lipids from pressed palm oil (Elaes guineensis) fibers using supercritical CO_2. *The Journal of Supercritical Fluids* 18 (1): 35–47.

Franca, L.F., Reber, G., Meireless, M.A.A. et al. (1999). Supercritical extraction of carotenoids and lipids from buriti (mauritia flexuosa), a fruit from the Amazon region. *Journal of Supercritical Fluids* 14: 247–256.

Funazukuri, T., Kong, C., and Kagei, S. (1998). Effective axial dispersion coefficients in packed beds under supercritical conditions. *The Journal of Supercritical Fluids* 13 (1–3): 169–175.

Garrigós, M., Reche, F., Pernas, K., and Jiménez, A. (2000). Optimization of parameters for the analysis of aromatic amines in finger-paints. *Journal of Chromatography A 896* (1-2): 291–298.

Gast, K., Jungfer, M., and Brunner, G. (2001). Enrichment of vitamin E and provitamin a from crude palm oil by supercritical fluid extraction. Paper presented at the Proceedings of the 2nd International Meeting on High Pressure Chemical Engineering (7–9 March 2001).

Geana, D. and Steiner, R. (1995). Calculation of phase equilibrium in supercritical extraction of C54 triglyceride (rapeseed oil). *The Journal of Supercritical Fluids 8* (2): 107–118.

Ghafoor, K., Park, J., and Choi, Y.-H. (2010). Optimization of supercritical fluid extraction of bioactive compounds from grape (Vitis labrusca B.) peel by using response surface methodology. *Innovative Food Science & Emerging Technologies 11* (3): 485–490.

Ghasemi, E., Raofie, F., and Najafi, N.M. (2011). Application of response surface methodology and central composite design for the optimisation of supercritical fluid extraction of essential oils from Myrtus communis L. leaves. *Food Chemistry 126* (3): 1449–1453.

Giovanni, M. (1983). Response surface methodology and product optimization. *Food Technology 37* (11): 41–45.

Goh, S., Choo, Y.M., and Ong, A.S. (1987). Minor components in palm oil. Paper presented at the Proceedings of the 1987 International Oil Palm/Palm Oil Conference. Progress and Prospects. Conference II. Technology, Kuala Lumpur, Malaysia (29 June–1 July 1987).

Gonçalves, M., Vasconcelos, A., de Azevedo, E.G. et al. (1991). On the application of supercritical fluid extraction to the deacidification of olive oils. *Journal of the American Oil Chemists Society 68* (7): 474–480.

Gonçalves, C.B., Rodrigues, C.E., Marcon, E.C., and Meirelles, A.J. (2016). Deacidification of palm oil by solvent extraction. *Separation and Purification Technology 160*: 106–111.

Goodrum, J.W. and Kilgo, M.B. (1987). Peanut oil extraction with SC-CO_2: solubility and kinetic functions. *Transactions of ASAE 30*: 1865–1868.

Goto, M., Sato, M., and Hirose, T. (1993). Extraction of peppermint oil by supercritical carbon dioxide. *Journal of Chemical Engineering of Japan 26* (4): 401–407.

Griffith, K.N. (2001). Environmentally Benign Chemical Processing in Near- and Supercritical Fluids and Gases Expanded Liquids. Thesis for Degree Doctor of Philosophy in Chemistry. Georgia Institute Technology.

Gülcin, I., Oktay, M., Kıreçcı, E., and Küfrevıoğlu, Ö.I. (2003). Screening of antioxidant and antimicrobial activities of anise (Pimpinella anisum L.) seed extracts. *Food Chemistry 83*: 371–382.

Gunstoner, F.D. (1987). *Palm Oil*. UK: Society of Chemical Industry.

Guo, B., Li, D., Cheng, C. et al. (2001). Simulation of biomass gasification with a hybrid neural network model. *Bioresource Technology* 76 (2): 77–83.

Hagan, M., Demuth, H., and Beale, M. (1995). *Neural Network Design* (eds. J. Andreae and D. Foresee), 2–23. Boston, MA: PWS Publishing Co.

Han, Y., Ma, Q., Wang, L., and Xue, C. (2012). Extraction of astaxanthin from Euphausia pacific using subcritical 1, 1, 1, 2-tetrafluoroethane. *Journal of Ocean University of China* 11 (4): 562–568.

Hansen, B.N., Harvey, A.H., Coelho, J.A.P. et al. (2001). Solubility of capsaicin and β-carotene in supercritical carbon dioxide and in halocarbons. *Journal of Chemical & Engineering Data* 46 (5): 1054–1058.

Hawthorne, S.B. and King, J.W. (1999). Principles and practice of analytical SFE. In: *Practical Supercritical Fluid Chromatography and Extraction* (eds. M. Caude and D. Thiebaut), 219–282. Amsterdam: Harwood Academic.

He, L., Zhang, X., Xu, H. et al. (2012). Subcritical water extraction of phenolic compounds from pomegranate (Punica granatum L.) seed residues and investigation into their antioxidant activities with HPLC–ABTS+ assay. *Food and Bioproducts Processing* 90 (2): 215–223.

Heaton, J. (2015). *AIFH, Volume 3: Deep Learning and Neural Networks*. Chesterfield, MO: Heaton Research, Inc.

Henderson, D., Jacobson, S.H., and Johnson, A.W. (2003). The theory and practice of simulated annealing. In: *Handbook of Metaheuristics* (eds. F. Glover and G.A. Kochenberger), 287–319. Springer.

Herrero, M., Cifuentes, A., and Ibañez, E. (2006). Sub-and supercritical fluid extraction of functional ingredients from different natural sources: plants, food-by-products, algae and microalgae: a review. *Food Chemistry* 98 (1): 136–148.

Hocaoglu, F.O., Oysal, Y., and Kurban, M. (2009). Missing wind data forecasting with adaptive neuro-fuzzy inference system. *Neural Computing and Applications* 18 (3): 207–212.

Holland, J. (1975). *Adaptation in Natural and Artificial Systems: An Introductory Analysis with Applications to Biology, Control, and Artificial Intelligence*. Oxford, England: University of Michigan Press.

Hong, I., Rho, S., Lee, K. et al. (1990). Modeling of soybean oil bed extraction with supercritical carbon dioxide. *Korean Journal of Chemical Engineering* 7 (1): 40–46.

Hu, A.-j., Zhao, S., Liang, H. et al. (2007). Ultrasound assisted supercritical fluid extraction of oil and coixenolide from adlay seed. *Ultrasonics Sonochemistry* 14 (2): 219–224.

Hubert, P. and Vitzthum, O.G. (1978). Fluid extraction of hops, spices, and tobacco with supercritical gases. *Angewandte Chemie International Edition in English* 17 (10): 710–715.

Hui, Y. (1996). Bailey's industrial oil and fat products. In: *Edible Oil and Fat ProductsOils and Oil Seeds*, vol. *2*. Wiley.

Hurtado-Benavides, A.M., Señoráns, F.J., Ibáñez, E., and Reglero, G. (2004). Countercurrent packed column supercritical CO_2 extraction of olive oil. Mass transfer evaluation. *The Journal of Supercritical Fluids* 28 (1): 29–35.

Illés, V., Daood, H.G., Perneczki, S. et al. (2000). Extraction of coriander seed oil by SC-CO_2 and propane at super- and subcritical conditions. *Journal of Supercritical Fluids* 17: 177–186.

Ivanovic, J., Ristic, M., and Skala, D. (2011). Supercritical CO_2 extraction of Helichrysum italicum: influence of CO_2 density and moisture content of plant material. *The Journal of Supercritical Fluids* 57 (2): 129–136.

Ivanovic, J., Meyer, F., Stamenic, M. et al. (2014). Pretreatment of natural materials used for supercritical fluid extraction of commercial phytopharmaceuticals. *Chemical Engineering & Technology* 37 (9): 1606–1611.

Jang, J.-S. (1993). ANFIS: adaptive-network-based fuzzy inference system. *IEEE Transactions on Systems, Man, and Cybernetics* 23 (3): 665–685.

Jang, J.-S. and Sun, C.-T. (1995). Neuro-fuzzy modeling and control. *Proceedings of the IEEE* 83 (3): 378–406.

Jang, J.-S.R., Sun, C.-T., and Mizutani, E. (1997). Neuro-fuzzy and soft computing-a computational approach to learning and machine intelligence [Book Review]. *IEEE Transactions on Automatic Control* 42 (10): 1482–1484.

Japir, A.A.-W., Salimon, J., Derawi, D. et al. (2017). Physicochemical characteristics of high free fatty acid crude palm oil. *Oilseeds and Fats, Crops and Lipids* 24 (5): D506.

Johannsen, M. and Brunner, G. (1997). Solubilities of the fat-soluble vitamins A, D, E, and K in supercritical carbon dioxide. *Journal of Chemical & Engineering Data* 42: 106–111.

Kalra, H., Chung, S.Y.K., and Chen, C.J. (1987). Phase equilibrium data for supercritical extraction of lemon flavors and palm oils with carbon dioxide. *Fluid Phase Equilibria* 36: 263–278.

Katz, S.N. (1989). Method for decaffeinating coffee with a supercritical fluid. Patent 4820537. filed 8 March 1988.

Khaw, K.Y., Parat, M.O., Shaw, P.N., and Falconer, J.R. (2017). Solvent supercritical fluid technologies to extract bioactive compounds from natural sources: a review. *Molecules* 22 (7): 1186.

King, J.W. (1997). Critical fluids for oil extraction. In: *Technology and Solvents for Extracting Oilseeds and Nonpetroleum Oils*, 287–290. Champaign, IL: AOCS Press.

King, M. and Bott, T. (1995). Compressed and liquefied gases as solvents: the commercial applications. In: *Extraction of Natural Products Using Near-Critical*

Solvents (eds. M.B. King and T.R. Bott), 19–23. Glasgow: Blackie Academic & Professional.

King, M. and Catchpole, O. (1993). Physico-chemical data required for the design of near-critical fluid extraction process. In: *Extraction of Natural Products Using Near-Critical Solvents*, 184–231. Springer.

Kiriamiti, H., Rascol, E., Marty, A., and Condoret, J. (2002). Extraction rates of oil from high oleic sunflower seeds with supercritical carbon dioxide. *Chemical Engineering and Processing: Process Intensification 41* (8): 711–718.

Kiselev, S., Perkins, R.A., and Huber, M.L. (1999). Transport properties of refrigerants R32, R125, R134a, and R125+ R32 mixtures in and beyond the critical region. *International Journal of Refrigeration 22* (6): 509–520.

Klein, J. (1981). Physical and chemical refining of soybean oil [Glycine max]. *Revue Francaise des Corps Gras 28*: 301–313.

Klein, T. and Schulz, S. (1989). Measurement and model prediction of vapor-liquid equilibria of mixtures of rapeseed oil and supercritical carbon dioxide. *Industrial & Engineering Chemistry Research 28* (7): 1073–1081.

Knez, Ž., Škerget, M., and Uzunalić, A.P. (2007). Phase equilibria of vanillins in compressed gases. *The Journal of Supercritical Fluids 43* (2): 237–248.

Koseoglu, S.S., Rhee, K.C., and Wilson, R.F.E. (1996). *An Overview of Supercritical Fluid Extraction*. Champaign, IL: AOCS Press.

Koseoglu, S.S., Rhee, K.C., and Wilson, R.F.E. (1998). Processing of oilseed constituents for non-food applications. In: *Emerging Technologies, Current Practices, Quality Control, Technology Transfer, and Environmental Issues*, vol. *1* (eds. J.T.P. Derksen and F.P. Cuperus), 113–120. Champaign, IL: AOCS Press.

Kostrzewa, D., Dobrzyńska-Inger, A., and Reszczyński, R. (2020). Pilot scale supercritical CO_2 extraction of carotenoids from sweet paprika (Capsicum annuum L.): influence of particle size and moisture content of plant material. *LWT – Food Science and Technologys 136*: 110345.

Kotnik, P., Škerget, M., and Knez, Ž. (2007). Supercritical fluid extraction of chamomile flower heads: comparison with conventional extraction, kinetics and scale-up. *The Journal of Supercritical Fluids 43* (2): 192–198.

Koza, J.R. and Koza, J.R. (1992). *Genetic Programming: On the Programming of Computers by Means of Natural Selection*, vol. *1*. MIT Press.

Kraujalis, P. and Venskutonis, P.R. (2013). Supercritical carbon dioxide extraction of squalene and tocopherols from amaranth and assessment of extracts antioxidant activity. *The Journal of Supercritical Fluids 80*: 78–85.

Krauss, R., Luettmer-Strathmann, J., Sengers, J.V., and Stephan, K. (1993). Transport properties of 1,1,1,2-tetrafluoroethane (R134a). *International Journal of Thermophysics 14* (4): 951–988.

Krishnamurthy, R., Widlak, N.R., and Wang, J.J. (1992). Methods for treatment of edible oils. Patent 5091116, filed 26 November 1986.

Kumar, P.P. and Krishna, A.G. (2014). Physico-chemical characteristics and nutraceutical distribution of crude palm oil and its fractions. *Grasas y Aceites 65* (2): 018.

Kumar, B.S. and Venkateswarlu, C. (2012). Estimating biofilm reaction kinetics using hybrid mechanistic-neural network rate function model. *Bioresource Technology 103* (1): 300–308.

Lagalante, A.F., Clarke, A.M., and Bruno, T.J. (1998). Modeling the water-R134a partition coefficients of organic solutes using a linear solvation energy relationship. *The Journal of Physical Chemistry B 102* (44): 8889–8892.

Lang, R. (2001). *A Future for Dynamic Neural Networks*. UK: Dept. Cybernetics, University of Reading.

Lang, Q. and Wai, C.M. (2001). Supercritical fluid extraction in herbal and natural product studies–a practical review. *Talanta 53* (4): 771–782.

Lau, H.L.N., Choo, Y.M., Ma, A.N., and Chuah, C.H. (2006). Characterization and supercritical carbon dioxide extraction of palm oil (Elaeis guineensis). *Journal of Food Lipids 13* (2): 210–221.

Lau, H.L.N., Choo, Y.M., Ma, A.N., and Chuah, C.H. (2008). Selective extraction of palm carotene and vitamin E from fresh palm-pressed mesocarp fiber (Elaeis guineensis) using supercritical CO_2. *Journal of Food Engineering 84* (2): 289–296.

Lee, A., Bulley, N., Fattori, M., and Meisen, A. (1986). Modelling of supercritical carbon dioxide extraction of canola oilseed in fixed beds. *Journal of the American Oil Chemists' Society 63* (7): 921–925.

Leeke, G., Gaspar, F., and Santos, R. (2002). Influence of water on the extraction of essential oils from a model herb using supercritical carbon dioxide. *Industrial & Engineering Chemistry Research 41* (8): 2033–2039.

Li, B., Xu, Y., Jin, Y.-X. et al. (2010). Response surface optimization of supercritical fluid extraction of kaempferol glycosides from tea seed cake. *Industrial Crops and Products 32* (2): 123–128.

Li, Y., Fabiano-Tixier, A.S., Vian, M.A., and Chemat, F. (2013). Solvent-free microwave extraction of bioactive compounds provides a tool for green analytical chemistry. *TrAC Trends in Analytical Chemistry 47*: 1–11.

Lim, C.S., Manan, Z.A., and Sarmidi, M.R. (2003). Simulation modeling of the phase behavior of palm oil-supercritical carbon dioxide. *Journal of the American Oil Chemists' Society 80* (11): 1147–1156.

List, G., King, J., Johnson, J. et al. (1993). Supercritical CO_2 degumming and physical refining of soybean oil. *Journal of the American Oil Chemists Society 70* (5): 473–476.

Liu, S., Yang, F., Zhang, C. et al. (2009). Optimization of process parameters for supercritical carbon dioxide extraction of Passiflora seed oil by response surface methodology. *The Journal of Supercritical Fluids 48* (1): 9–14.

Liu, X., Ou, H., Xiang, Z., and Gregersen, H. (2020). Ultrasound pretreatment combined with supercritical CO_2 extraction of Iberis amara seed oil. *Journal of Applied Research on Medicinal and Aromatic Plants 18*: 100265.

López-Padilla, A., Ruiz-Rodriguez, A., Reglero, G., and Fornari, T. (2017). Supercritical carbon dioxide extraction of Calendula officinalis: kinetic modeling and scaling up study. *Journal of Supercritical Fluids 130*: 292–300.

Lorenzo, R., Vazquez, M., Carro, A., and Cela, R. (1999). Methylmercury extraction from aquatic sediments: a comparison between manual, supercritical fluid and microwave-assisted techniques. *TrAC Trends in Analytical Chemistry 18* (6): 410–416.

Lu, J., Feng, X., Han, Y., and Xue, C. (2014). Optimization of subcritical fluid extraction of carotenoids and chlorophyll a from Laminaria japonica Aresch by response surface methodology. *Journal of the Science of Food and Agriculture 94* (1): 139–145.

Machado, N.T. and Brunner, G. (1997). High pressure vapor-liquid equilibria of palm fatty acids distillates-carbon dioxide system. *Ciência e Tecnologia de Alimentos 17* (4): 354–360.

Machmudah, S., Sulaswatty, A., Sasaki, M. et al. (2006). Supercritical CO_2 extraction of nutmeg oil: experiments and modeling. *The Journal of Supercritical Fluids 39* (1): 30–39.

Machmudah, S., Kawahito, Y., Sasaki, M., and Goto, M. (2007). Supercritical CO_2 extraction of rosehip seed oil: fatty acids composition and process optimization. *The Journal of Supercritical Fluids 41* (3): 421–428.

Maclellan, M. (1983). Palm oil. *Journal of the American Oil Chemists' Society 60* (2): 368–378.

Mag, T.K. (1994). Changes in Canada's fats and oils industry. *Inform 5* (7): 827–832.

Maheshwari, P., Nikolov, Z., White, T., and Hartel, R. (1992). Solubility of fatty acids in supercritical carbon dioxide. *Journal of the American Oil Chemists' Society 69* (11): 1069–1076.

Mamdani, E.H. and Assilian, S. (1975). An experiment in linguistic synthesis with a fuzzy logic controller. *International Journal of Man-Machine Studies 7* (1): 1–13.

Manan, Z.A., Siang, L.C., and Mustapa, A.N. (2009). Development of a new process for palm oil refining based on supercritical fluid extraction technology. *Industrial & Engineering Chemistry Research 48* (11): 5420–5426.

Marcelo, M.M., Barbosa, H.M., Passos, C.P., and Silva, C.M. (2014). Supercritical fluid extraction of spent coffee grounds: measurement of extraction curves, oil characterization and economic analysis. *The Journal of Supercritical Fluids* 86: 150–159.

Mariano, R.G.d.B., Couri, S., and Freitas, S.P. (2009). Enzymatic technology to improve oil extraction from Caryocar brasiliense camb.(Pequi) Pulp. *Revista Brasileira de Fruticultura 31* (3): 637–643.

Markom, M., Singh, H., and Hasan, M. (1999). Solubility of palm oil in supercritical CO_2 in comparison with other edible oils. Proceedings of World Engineering Congress 1999, UPM Press, 23–28.

Mathias, P.M. (1983). A versatile phase equilibrium equation of state. *Industrial and Engineering Chemistry Process Design and Development 22* (3): 385–391.

Mathias, P.M., Copeman, T.W., and Prausnitz, J.M. (1986). Phase equilibria for supercritical extraction of lemon flavors and palm oils with carbon dioxide. *Fluid Phase Equilibria 29*: 545–554.

MATLAB (2004). MATLAB, The Language of Technical Computing. The Math Works.

McCulloch, A. (1999). CFC and Halon replacements in the environment. *Journal of Fluorine Chemistry 100* (1-2): 163–173.

McHugh, M.A. and Krukonis, V. (1994). *Supercritical Fluid Extraction: Principles and Practice*. Stoneham, MA: Butterworth-Heinemann.

Meier, U., Gross, F., and Trepp, C. (1994). High-pressure phase-equilibrium studies for the carbon-dioxide alpha-tocopherol (vitamin-E) system. *Fluid Phase Equilibria 92*: 289–302.

de Melo, S.V., Costa, G.M.N., Viana, A., and Pessoa, F. (2009). Solid pure component property effects on modeling upper crossover pressure for supercritical fluid process synthesis: a case study for the separation of Annatto pigments using SC-CO_2. *The Journal of Supercritical Fluids 49* (1): 1–8.

Melreles, M., Zahedi, G., and Hatami, T. (2009). Mathematical modeling of supercritical fluid extraction for obtaining extracts from vetiver root. *The Journal of Supercritical Fluids 49* (1): 23–31.

Mendes, M.F., Pessoa, F.L.P., and Uller, A.M.C. (2002). An economic evaluation based on an experimental study of the vitamin E concentration present in deodorizer distillate of soybean oil using supercritical CO_2. *The Journal of Supercritical Fluids 23* (3): 257–265.

Mendes, R.L., Reis, A.D., Pereira, A.P. et al. (2005). Supercritical CO_2 extraction of γ-linolenic acid (GLA) from the cyanobacterium Arthrospira (Spirulina) maxima: experiments and modeling. *Chemical Engineering Journal 105* (3): 147–151.

Michell, M. (1998). *An Introduction to Genetic Algorithms [Electronic Resource].* MIT Press.

Michielin, E.M., Salvador, A.A., Riehl, C.A. et al. (2009). Chemical composition and antibacterial activity of Cordia verbenacea extracts obtained by different methods. *Bioresource Technology 100* (24): 6615–6623.

Mongkholkhajornsilp, D., Doulas, S., Douglas, P. et al. (2005). Supercritical CO extraction of nimbin from neem seeds – a modeling study. *Journal of Food Engineering 71*: 331–340.

MPOPC (2002). Malaysian Palm Oil Industry Overview: 2001. *Palm Oil Link 12* (1): 1–3.

Mustapa, A., Manan, Z., Azizi, C.M. et al. (2009). Effects of parameters on yield for sub-critical R134a extraction of palm oil. *Journal of Food Engineering 95* (4): 606–616.

Mustapa, A., Manan, Z., Azizi, C.M. et al. (2011). Extraction of β-carotenes from palm oil mesocarp using sub-critical R134a. *Food Chemistry 125* (1): 262–267.

Mustapa, A.N., Martin, Á., Mato, R.B., and Cocero, M.J. (2015). Extraction of phytocompounds from the medicinal plant Clinacanthus nutans Lindau by microwave-assisted extraction and supercritical carbon dioxide extraction. *Industrial Crops and Products 74*: 83–94.

Nagy, B. and Simándi, B. (2008). Effects of particle size distribution, moisture content, and initial oil content on the supercritical fluid extraction of paprika. *Journal of Supercritical Fluids 46* (3): 293–298.

Nehlig, A. (1999). Are we dependent upon coffee and caffeine? A review on human and animal data. *Neuroscience & Biobehavioral Reviews 23* (4): 563–576.

Norhuda, I. (2005). Studies on mass transfer characteristics of palm kernel oil extraction using supercritical carbon dioxide extraction. PhD thesis. Universiti Sains Malaysia.

Norulaini, N.N., Anuar, O., Abbas, F. et al. (2009). Optimization of supercritical CO_2 extraction of Anastatica hierochuntica. *Food and Bioproducts Processing 87* (2): 152–158.

Nyam, K.L., Tan, C.P., Karim, R. et al. (2010). Extraction of tocopherol-enriched oils from Kalahari melon and roselle seeds by supercritical fluid extraction (SFE-CO_2). *Food Chemistry 119* (3): 1278–1283.

Oghaki, K. (1989). A fundamental study of the extraction with a supercritical fluid Solubilities of α-tocopherol palmitic acid and tripalmitin in compressed carbon dioxide at 25 and 40°C. *International Chemical Engineering Japan 29*: 303.

Oghaki, K., Tsukahara, I., Semba, K., and Katayama, T. (1989). A fundamental study of the extraction with a supercritical fluid Solubilities of α-tocopherol palmitic acid and tripalmitin in compressed carbon dioxide at 25 and 40°C. *International Chemical Engineering of Japan 29* (2): 302–308.

Ohgaki, K. and Katayama, T. (1992). Extractive separation method. Patent 5138075, filed 1 September 1989.

Oo, K.-c., Lee, K.-B., and Ong, A.S. (1986). Changes in fatty acid composition of the lipid classes in developing oil palm mesocarp. *Phytochemistry 25* (2): 405–407.

Ooi, C., Bhaskar, A., Yener, M. et al. (1996). Continuous supercritical carbon dioxide processing of palm oil. *Journal of the American Oil Chemists Society 73* (2): 233–237.

Özkal, S.G., Salgın, U., and Yener, M. (2005). Supercritical carbon dioxide extraction of hazelnut oil. *Journal of Food Engineering 69* (2): 217–223.

Papamichail, I., Louli, V., and Magoulas, K. (2000). Supercritical fluid extraction of celery seed oil. *The Journal of Supercritical Fluids 18* (3): 213–226.

Passos, C.P., Silva, R.M., Da Silva, F.A. et al. (2009). Enhancement of the supercritical fluid extraction of grape seed oil by using enzymatically pre-treated seed. *The Journal of Supercritical Fluids 48* (3): 225–229.

Peng, D.-Y. and Robinson, D.B. (1976). A new two-constant equation of state. *Industrial & Engineering Chemistry Fundamentals 15* (1): 59–64.

Pereira, P.J., Goncalves, M., Coto, B. et al. (1993). Phase equilibria of CO_2+ dl-α-tocopherol at temperatures from 292 K to 333 K and pressures up to 26 MPa. *Fluid Phase Equilibria 91* (1): 133–143.

Pereira, C.G., Leal, P.F., Sato, D.N., and Meireles, M.A.A. (2005). Antioxidant and antimycobacterial activities of Tabernaemontana catharinensis extracts obtained by supercritical CO_2+ cosolvent. *Journal of Medicinal Food 8* (4): 533–538.

Perry, E.S., Weber, W.H., and Daubert, B.F. (1949). Vapour pressures of phlegmatic liquids, I. Simple and mixed triglycerides. *Journal of the American Chemical Society 71*: 3720–3726.

Pietsch, A. and Eggers, R. (1999). The mixer-settler principle as a separation unit in supercritical fluid processes. *Journal of Supercritical Fluids 14* (2): 163–171.

Pimentel, F.A., Cardoso, M.d.G., Guimarães, L.G. et al. (2013). Extracts from the leaves of Piper piscatorum (Trel. Yunc.) obtained by supercritical extraction of with CO_2, employing ethanol and methanol as co-solvents. *Industrial Crops and Products 43*: 490–495.

Potter, C.W. and Negnevitsky, M. (2006). Very short-term wind forecasting for Tasmanian power generation. *IEEE Transactions on Power Systems 21* (2): 965–972.

Pouralinazar, F., Yunus, M.A.C., and Zahedi, G. (2012). Pressurized liquid extraction of Orthosiphon stamineus oil: experimental and modeling studies. *The Journal of Supercritical Fluids* 62: 88–95.

Povh, N.P., Marques, M.O., and Meireles, M.A.A. (2001). Supercritical CO_2 extraction of essential oil and oleoresin from chamomile (Chamomilla recutita [L.] Rauschert). *The Journal of Supercritical Fluids* 21 (3): 245–256.

Psichogios, D.C. and Ungar, L.H. (1992). A hybrid neural network-first principles approach to process modeling. *AICHE Journal* 38 (10): 1499–1511.

Quah, J.T. and Ng, W. (2007). Utilizing computational intelligence for DJIA stock selection. Paper presented at the 2007 International Joint Conference on Neural Networks (12–17 August 2007).

Quispe-Condori, S., Sánchez, D., Foglio, M.A. et al. (2005). Global yield isotherms and kinetic of artemisinin extraction from Artemisia annua L leaves using supercritical carbon dioxide. *The Journal of Supercritical Fluids* 36 (1): 40–48.

Quitain, A.T., Moriyoshi, T., and Goto, M. (2013). Coupling microwave-assisted drying and supercritical carbon dioxide extraction for coconut oil processing. *Chemical Engineering Science 1* (1): 12–16.

Rahimi, E., Prado, J., Zahedi, G., and Meireles, M. (2011). Chamomile extraction with supercritical carbon dioxide: Mathematical modeling and optimization. *The Journal of Supercritical Fluids* 56 (1): 80–88.

Raman, L., Cheryan, M., and Rajagopalan, N. (1996). Deacidification of soybean oil by membrane technology. *Journal of the American Oil Chemists' Society* 73 (2): 219–224.

Reid, C., Prausnitz, J.M., and Poling, B.E. (1987). *The Properties of Gases and Liquids*, 4the. New York: McGraw-Hill.

Renner, G. and Ekárt, A. (2003). Genetic algorithms in computer aided design. *Computer-Aided Design* 35 (8): 709–726.

Reverchon, E. (1996). Mathematical modeling of supercritical extraction of sage oil. *AICHE Journal* 42 (6): 1765–1771.

Reverchon, E. (1997). Supercritical fluid extraction and fractionation of essential oils and related products. *The Journal of Supercritical Fluids* 10 (1): 1–37.

Reverchon, E. and De Marco, I. (2006). Supercritical fluid extraction and fractionation of natural matter. *The Journal of Supercritical Fluids* 38 (2): 146–166.

Reverchon, E. and Osseo, L.S. (1994). Comparison of processes for the supercritical carbon dioxide extraction of oil from soybean seeds. *Journal of the American Oil Chemists' Society* 71: 1007–1012.

Rivizi, S., Daniels, J., Benado, A., and Zollweg, J. (1986). Supercritical fluid extraction operation principles and food application. *Food Technology* 40: 57–64.

Rizvi, S.S.H., Yu, Z. R., Bhaskar, A. R. and Chidambara Raj, C. B. (1994). Fundamental of Processing with Supercritical Fluids. In: Rizvi, S. S. H. (Ed) Supercritical Fluid Processing of Food and Biomaterials, Blackie Academic and Professional. Glasgow NZ, Chapman & Hall, (pp. 1–26).

Roth, M. (1996). Thermodynamic prospects of alternative refrigerants as solvents for supercritical fluid extraction. *Analytical Chemistry 68* (24): 4474–4480.

Ruivo, R., Cebola, M.J., Simões, P.C., and Nunes da Ponte, M. (2001). Fractionation of edible oil model mixtures by supercritical carbon dioxide in a packed column. Part I: experimental results. *Industrial and Engineering Chemistry Research 40*: 1706–1711.

Rüütmann, T. and Kallas, J. (1994). Deodorization of cocoa butter, fish fat and the fat of fur-bearing animals. *Fett (Germany) 96*: 259–266.

Sahashi, Y., Ishizuka, H., Koike, S., and Suzuki, K. (1994). Membrane separation of polyunsaturated fatty acids-rich acylglycerols from fish oil. *Journal of Japan Oil Chemists' Society 43* (2): 116–123.

Saldaña, M.D., Mohamed, R.S., and Mazzafera, P. (2002). Extraction of cocoa butter from Brazilian cocoa beans using supercritical CO_2 and ethane. *Fluid Phase Equilibria 194*: 885–894.

Salgın, U. and Salgın, S. (2013). Effect of main process parameters on extraction of pine kernel lipid using supercritical green solvents: solubility models and lipid profiles. *The Journal of Supercritical Fluids 73*: 18–27.

Salgın, U., Döker, O., and Çalımlı, A. (2006). Extraction of sunflower oil with supercritical CO_2: experiments and modeling. *The Journal of Supercritical Fluids 38* (3): 326–331.

Sambanthamurthi, R., Sundram, K., and Tan, Y.-A. (2000). Chemistry and biochemistry of palm oil. *Progress in Lipid Research 39* (6): 506–558.

Sandler, S.I. (1994). *Models for Thermodynamic and Phase Equilibria Calculations*. New York: Marcel Dekker.

Santos, P., Aguiar, A.C., Barbero, G.F. et al. (2015). Supercritical carbon dioxide extraction of capsaicinoids from malagueta pepper (Capsicum frutescens L.) assisted by ultrasound. *Ultrasonics Sonochemistry 22*: 78–88.

Senorans, F., Ibanez, E., Cavero, S. et al. (2000). Liquid chromatographic–mass spectrometric analysis of supercritical-fluid extracts of rosemary plants. *Journal of Chromatography A 870* (1–2): 491–499.

Shacham, M., Brauner, N., and Citlip, M. (1995). Critical analysis of experimental data, regression models and regression coefficients in data correlation. Paper presented at the AIChE Symposium Series (12–17 November 1995).

Sharif, K., Rahman, M., Azmir, J. et al. (2014). Experimental design of supercritical fluid extraction – a review. *Journal of Food Engineering 124*: 105–116.

Sheikhattar, L., Hashim, H., and Zahedi, G. (2011). Artificial neural network simulation and sensitivity analysis of heavy oil cracking unit. *International Journal of Chemical and Environmental Engineering* 2 (1): 7–14.

Shi, B., Jin, J., Yu, E., and Zhang, Z. (2011). Concentration of natural vitamin E using a continuous countercurrent supercritical CO_2 extraction-distillation dual column. *Chemical Engineering & Technology* 34 (6): 914–920.

Shokri, A., Hatami, T., and Khamforoush, M. (2011). Near critical carbon dioxide extraction of Anise (Pimpinella Anisum L.) seed: mathematical and artificial neural network modeling. *The Journal of Supercritical Fluids* 58 (1): 49–57.

Shucheng, L., Feng, Y., Chaohua, Z. et al. (2009). Optimization of process parameters for supercritical carbon dioxide extraction of Passiflora seed oil by response surface methodology. *The Journal of Supercritical Fluids* 48 (1): 9–14.

Simoes, P.C. and Catchpole, O.J. (2002). Fractionation of lipid mixtures by subcritical R134a in a packed column. *Industrial & Engineering Chemistry Research* 41 (2): 267–276.

Singh, P.P. and Saldaña, M.D. (2011). Subcritical water extraction of phenolic compounds from potato peel. *Food Research International* 44 (8): 2452–2458.

Škerget, M., Kotnik, P., and Knez, Ž. (2002). Phase equilibria in systems containing α-tocopherol and dense gas. *Journal of Supercritical Fluids* 26 (3): 181–191.

Snyder, J., Friedrich, J., and Christianson, D. (1984). Effect of moisture and particle size on the extractability of oils from seeds with supercritical CO_2. *Journal of the American Oil Chemists Society* 61 (12): 1851–1856.

Soave, G. (1972). Equilibrium constants from a modified Redlich-Kwong equation of state. *Chemical Engineering Science* 27 (6): 1197–1203.

Solana, M., Boschiero, I., Dall'Acqua, S., and Bertucco, A. (2014). Extraction of bioactive enriched fractions from Eruca sativa leaves by supercritical CO_2 technology using different co-solvents. *The Journal of Supercritical Fluids* 94: 245–251.

Sovová, H. (1994). Rate of the vegetable oil extraction with supercritical CO_2–I. Modelling of extraction curves. *Chemical Engineering Science* 49 (3): 409–414.

Sovová, H. (2005). Mathematical model for supercritical fluid extraction of natural products and extraction curve evaluation. *The Journal of Supercritical Fluids* 33 (1): 35–52.

Sovová, H. (2012). Steps of supercritical fluid extraction of natural products and their characteristic times. *The Journal of Supercritical Fluids* 66: 73–79.

Sovová, H., Stateva, R.P., and Galushko, A.A. (2001). Essential oils from seeds: solubility of limonene in supercritical CO_2 and how it is affected by fatty oil. *Journal of Supercritical Fluids* 20 (2): 113–129.

Sporring, S., Bøwadt, S., Svensmark, B., and Björklund, E. (2005). Comprehensive comparison of classic Soxhlet extraction with Soxtec extraction, ultrasonication extraction, supercritical fluid extraction, microwave assisted extraction and accelerated solvent extraction for the determination of polychlorinated biphenyls in soil. *Journal of Chromatography A 1090* (1–2): 1–9.

Stahl, E., Quirin, K.W., and Gerard, D. (1986). *Verdichtete Gase zur Extraktion und Raffination (Condensed Gases for the Extraction and Refining)*. Berlin: Springler-Verlag.

Stahl, E., Quirin, K.-W., and Gerard, D. (1988). Applications of dense gases to extraction and refining. In: *Dense Gases for Extraction and Refining* (ed. M.R.F. Ashworth), 72–217. Springer.

Stamenic, M., Zizovic, I., Eggers, R. et al. (2010). Swelling of plant material in supercritical carbon dioxide. *The Journal of Supercritical Fluids 52* (1): 125–133.

Statista, 2021. https://www.statista.com/statistics/263933/production-of-vegetable-oils-worldwide-since-2000/ accessed on August 23, 2021.

Stoldt, J. and Brunner, G. (1998). Phase equilibrium measurements in complex systems of fats, fat compounds and supercritical carbon dioxide. *Fluid Phase Equilibria 146* (1-2): 269-295.

Stoldt, J. and Brunner, G. (1999). Phase equilibria in complex systems of palm oil deodorizer condensates and supercritical carbon dioxide: experiments and correlation. *Journal of Supercritical Fluids 14* (3): 181–195.

Su, T., Wang, W., Lv, Z. et al. (2016). Rapid Delaunay triangulation for randomly distributed point cloud data using adaptive Hilbert curve. *Computers & Graphics 54*: 65–74.

Subra, P., Castellani, S., Ksibi, H., and Garrabos, Y. (1997). Contribution to the determination of the solubility of β-carotene in supercritical carbon dioxide and nitrous oxide: experimental data and modeling. *Fluid Phase Equilibria 131* (1–2): 269–286.

Sugeno, M. and Kang, G. (1988). Structure identification of fuzzy model. *Fuzzy Sets and Systems 28* (1): 15–33.

Sui, X., Feng, X.M., Yue, R.Y. et al. (2014). Optimization of Subcritical 1, 1, 1, 2-tetrafluoroethane (R134a) removal of cholesterol from spray-dried Sthenoteuthis oualaniensis egg powder using response surface methodology. In: *Advanced Materials Research*, vol. *1033*, 717–723. Trans Tech Publications Ltd.

Sulihatimarsyila, A.N., Lau, H.L., Nabilah, K., and Azreena, I.N. (2019). Refining process for production of refined palm-pressed fibre oil. *Industrial Crops and Products 129*: 488–494.

Sun, M. and Temelli, F. (2006). Supercritical carbon dioxide extraction of carotenoids from carrot using canola oil as a continuous co-solvent. *Journal of Supercritical Fluids 37* (3): 397–408.

Sun, L., Rezaei, K.A., Temelli, F., and Ooraikul, B. (2002). Supercritical fluid extraction of alkylamides from Echinacea angustifolia. *Journal of Agricultural and Food Chemistry 50* (14): 3947–3953.

Sundram, K., Sambanthamurthi, R., and Tan, Y.-A. (2003). Palm fruit chemistry and nutrition. *Asia Pacific Journal of Clinical Nutrition 12* (3): 355–362.

Tahmasebi, P. and Hezarkhani, A. (2012). A hybrid neural networks-fuzzy logic-genetic algorithm for grade estimation. *Computers & Geosciences 42*: 18–27.

Takagi, T. and Sugeno, M. (1985). Fuzzy identification of systems and its applications to modeling and control. *IEEE Transactions on Systems, Man and Cybernetics 15* (1): 116–132.

Tan, B.K. and Oh, C.H. (1981). Malaysian palm oil chemical and physical characteristics. *PORIM Technology 3*: 2–3.

Tan, C., Liang, S., and Liou, D. (1988). Fluid-solid mass transfer in a supercritical fluid extractor. *Journal of Chemical Engineering 38*: 17–22.

Tatke, P. and Jaiswal, Y. (2011). An overview of microwave assisted extraction and its applications in herbal drug research. *Research Journal of Medicinal Plant 5* (1): 21–31.

Taylor, L.T. (1996). *Supercritical Fluid Extraction*. New York: Wiley.

Tegetmeier, A., Dittmar, D., Fredenhagen, A., and Eggers, R. (2000). Density and volume of water and triglyceride mixtures in contact with carbon dioxide. *Chemical Engineering Progress 39* (5): 399–405.

Temelli, F. (1992). Extraction of triglycerides and phospholipids from canola with supercritical carbon dioxide and ethanol. *Journal of Food Science 57* (2): 440–443.

TheMathWorksInc (2010). Fuzzy Logic ToolboxTM User's Guide. www.mathworks.com (accessed 8 May 2020).

Tilly, K.D., Chaplin, R.P., and Foster, N.R. (1990). Supercritical fluid extraction of the triglycerides present in vegetable oils. *Separation Science and Technology 25* (4): 357–367.

Toma, M., Vinatoru, M., Paniwnyk, L., and Mason, T.J. (2001). Investigation of the effects of ultrasound on vegetal tissues during solvent extraction. *Ultrasonics Sonochemistry 8* (2): 137–142.

Tonthubthimthong, P., Chuaprasert, S., Douglas, P., and Luewisutthichat, W. (2001). Supercritical CO_2 extraction of nimbin from neem seeds–an experimental study. *Journal of Food Engineering 47* (4): 289–293.

Treybal, R.E. (1990). *Mass Transfer Operations*, 3rde. New York: McGraw-Hill.

Turner, C., King, J.W., and Mathiasson, L. (2001). Supercritical fluid extraction and chromatography for fat-soluble vitamin analysis. *Journal of Chromatography A* 936 (1–2): 215–237.

Uno, K., Shibuta, Y., and Kataoka, Y. (1985). Purification of Tocopherol. Patent 0171009, filed 2 August 1984.

Uquiche, E., Jeréz, M., and Ortíz, J. (2008). Effect of pretreatment with microwaves on mechanical extraction yield and quality of vegetable oil from Chilean hazelnuts (Gevuina avellana Mol). *Innovative Food Science & Emerging Technologies* 9 (4): 495–500.

Üstündağ, Ö.G. and Temelli, F. (2000). Correlating the solubility behavior of fatty acids, mono-, di-, and triglycerides, and fatty acid esters in supercritical carbon dioxide. *Industrial and Engineering Chemistry Research* 39 (12): 4756–4766.

Van Broekhoven, E. and De Baets, B. (2009). Only smooth rule bases can generate monotone Mamdani–Assilian models under center-of-gravity defuzzification. *IEEE Transactions on Fuzzy Systems* 17 (5): 1157–1174.

Veggi, P., Martinez, J., and Meireles, M. (2013). Fundamentals of microwave extraction. In: *Microwave-assisted Extraction for Bioactive Compounds: Theory and Practice* (eds. F. Chemat and G. Cravotto). New York: Springer.

Vidović, S., Zeković, Z., Marošanović, B. et al. (2014). Influence of pre-treatments on yield, chemical composition and antioxidant activity of Satureja montana extracts obtained by supercritical carbon dioxide. *The Journal of Supercritical Fluids* 95: 468–473.

Wakao, N. and Funazkri, T. (1978). Effect of fluid dispersion coefficients on particle-to-fluid mass transfer coefficients in packed beds. *Chemical Engineering Science* 33: 1375–1384.

Walas, S.M. (1985). *Phase Equilibria in Chemical Engineering*. Butterworth-Heinemann.

Weast, R.C., Astle, M.J., and Beyer, W.H. (1990). *CRC Handbook of Chemistry and Physics*. Boca Raton, FL: CRC Press, Inc.

Weathers, R., Beckholt, D., Lavella, A., and Danielson, N. (1999). Comparison of acetals as in situ modifiers for the supercritical fluid extraction of β-carotene from paprika with carbon dioxide. *Journal of Liquid Chromatography & Related Technologies* 22 (2): 241–252.

Weber, W. and Brunner, G. (1995). Phase equilibria of triglycerides and gases. *AIChE Journal*, annual meeting, Chicago, IL, USA.

Weber, W., Petkov, S., and Brunner, G. (1999). Vapour-liquid-equilibria and calculations using the Redlich-Kwong-Aspen equation of state for tristearin, tripalmitin, and triolein in CO_2 and propane. *Fluid Phase Equilibria 158-160*: 695–706.

Wei, Z.-J., Liao, A.-M., Zhang, H.-X. et al. (2009). Optimization of supercritical carbon dioxide extraction of silkworm pupal oil applying the response surface methodology. *Bioresource Technology 100* (18): 4214–4219.

Wei, Y.-s., Zheng, M.-y., Geng, W., and Liu, J. (2012). Fatty acid composition analysis of common animal fats and vegetable oils [J]. *Food Science 16*: 188–193.

Wilde, P.F. (2000). Extraction of fixed and mineral oils with 1,1,1,2-tetrafluoroethane. UK patent 2,345,915, filed 5 March 1999.

Wilde, F.P. (2001). Extracting oil from a substance using iodotrifluoromethane. UK patent 2,352,724, filed 5 August 1999.

Wood, C.D. and Cooper, A.I. (2003). Synthesis of polystyrene by dispersion polymerization in 1, 1, 1, 2-tetrafluoroethane (R134a) using inexpensive hydrocarbon macromonomer stabilizers. *Macromolecules 36* (20): 7534–7542.

Wood, C.D., Senoo, K., Martin, C. et al. (2002). Polymer synthesis using hydrofluorocarbon solvents. 1. Synthesis of cross-linked polymers by dispersion polymerization in 1, 1, 1, 2-tetrafluoroethane. *Macromolecules 35* (18): 6743–6746.

Worden, K., Wong, C., Parlitz, U. et al. (2007). Identification of pre-sliding and sliding friction dynamics: grey box and black-box models. *Mechanical Systems and Signal Processing 21* (1): 514–534.

Xu, X., Gao, Y., Liu, G. et al. (2008). Optimization of supercritical carbon dioxide extraction of sea buckthorn (Hippophae thamnoides L.) oil using response surface methodology. *LWT- Food Science and Technology 41* (7): 1223–1231.

Xynos, N., Papaefstathiou, G., Gikas, E. et al. (2014). Design optimization study of the extraction of olive leaves performed with pressurized liquid extraction using response surface methodology. *Separation and Purification Technology 122*: 323–330.

Yang, C.T., Marsooli, R., and Aalami, M.T. (2009). Evaluation of total load sediment transport formulas using ANN. *International Journal of Sediment Research 24* (3): 274–286.

Yepez, B., Espinosa, M., López, S., and Bolanos, G. (2002). Producing antioxidant fractions from herbaceous matrices by supercritical fluid extraction. *Fluid Phase Equilibria 194*: 879–884.

You, J., Lao, W., and Wang, G. (1999). Analysis of organic pollutants in sewage by supercritical fluid extraction. *Chromatographia 49* (7-8): 399–405.

Yu, Z.R., Rizvi, S.S.H., and Zollweg, J.A. (1992). Phase equilibria of oleic acid, methyl oleate, and anhydrous milk fat in supercritical carbon dioxide. *Journal of Supercritical Fluids* 5 (2): 114–122.

Zadeh, L.A. (1965). Fuzzy sets. *Information and Control* 8 (3): 338–353.

Zadeh, L.A. (1973). Outline of a new approach to the analysis of complex systems and decision processes. *IEEE Transactions on Systems, Man, and Cybernetics* 1: 28–44.

Zahedi, G. and Azarpour, A. (2011). Optimization of supercritical carbon dioxide extraction of Passiflora seed oil. *The Journal of Supercritical Fluids* 58 (1): 40–48.

Zahedi, G., Elkamel, A., Lohi, A. et al. (2005). Hybrid artificial neural network–first principle model formulation for the unsteady state simulation and analysis of a packed bed reactor for CO_2 hydrogenation to methanol. *Chemical Engineering Journal* 115 (1–2): 113–120.

Zahedi, G., Elkamel, A., and Lohi, A. (2010a). Genetic algorithm optimization of supercritical fluid extraction of nimbin from neem seeds. *Journal of Food Engineering* 97 (2): 127–134.

Zahedi, G., Elkamel, A., Lohi, A., and Hatami, T. (2010b). Optimization of supercritical extraction of nimbin from neem seeds in presence of methanol as co-solvent. *The Journal of Supercritical Fluids* 55 (1): 142–148.

Zahedi, Y., Ghanbarzadeh, B., and Sedaghat, N. (2010c). Physical properties of edible emulsified films based on pistachio globulin protein and fatty acids. *Journal of Food Engineering* 100 (1): 102–108.

Zahedi, G., Elkamel, A., and Biglari, M. (2011). Optimization and sensitivity analysis of an extended distributed dynamic model of supercritical carbon dioxide extraction of nimbin from neem seeds. *Journal of Food Process Engineering* 34 (6): 2156–2176.

Zahedi, G., Azizi, S., Bahadori, A. et al. (2013). Electricity demand estimation using an adaptive neuro-fuzzy network: a case study from the Ontario province–Canada. *Energy* 49: 323–328.

Zargari, A. (1989). *Medicinal plants*, vol. 2. Tehran: Tehran University.

Zeković, Z., Vidović, S., Vladić, J. et al. (2014). Optimization of subcritical water extraction of antioxidants from Coriandrum sativum seeds by response surface methodology. *The Journal of Supercritical Fluids* 95: 560–566.

Ziegler, G.R. and Liaw, Y.J. (1993). Deodorization and deacidification of edible oils with dense carbon dioxide. *Journal of the American Oil Chemists' Society* 70 (10): 947–953.

Zizovic, I., Stamenić, M., Orlović, A., and Skala, D. (2007). Supercritical carbon dioxide extraction of essential oils from plants with secretory ducts:

mathematical modelling on the micro-scale. *The Journal of Supercritical Fluids* *39* (3): 338–346.

Zosel, K. (1974). Process for recovering caffeine. US3806619A, filed 7 May 1971.

Zosel, K. (1981). Process for the decaffeination of coffee. US4260639A, filed 28 January 1971.

Zosel, K. (1982). Process for the direct decaffeination of aqueous coffee extract solutions. Patent 4348422, filed 17 May 1978.

Index

a

acentric factor   109
alkali refining (chemical refining)   92
  bleaching   95
  degumming   93
  neutralisation   94
artificial intelligence
  fuzzy logic   61, 62, 66, 68
  neuro fuzzy   68, 69, 153–156, 158, 162
  transfer function   56–57, 63–65, 169, 172
Aspen Plus® 10.2.1 process simulator   102
  physical property database   107
  property constant estimation system   107
average absolute deviation (AAD)   112

b

base case process simulation
  palm oil deacidification process   142–145
  solubility of CPO in supercritical CO_2, 141–142
binary CPO component-CO_2 system properties
  temperature-dependent binary interaction parameters   122–132
  temperature-dependent polar factor   122–132
binary interaction parameters   110

c

chopping of palm fruits   200
coefficient of determination   187
coffee bean decaffeination   11
countercurrent extraction with external reflux   118, 213, 215

d

Deming algorithm   110
design of experiment
  full factorial   179
  multivariate   180
  response surface methodology (RSM)   165, 169, 170, 173–175, 180, 181, 228–231
diffusion-controlled phase   44

Index

e

empirical extraction efficiency   115
error   60, 72, 112, 153, 229–231
error
  mean square   153
  relative   229
  root mean square   168
  standard   187
extended Antoine equation   108
  correlated parameters   121
  generalized least squares regression   108
external mass transfer   39
extract phase   118, 145, 208
extraction yield
  formula   165, 171
  maximum   166
  optimum   168, 173

f

first principle model   49, 54, 232
flaking   200
flash module   117
  flash separator   117
  isothermal flash algorithm   132
free fatty acid (FFA)   16
free oil   42
fugacity coefficient   18

g

genetic algorithm
  optimum point   75, 223, 226

h

heterogeneity   198
high pressure
  cost   196
  economic   196

i

inert gas stripping   92
intraparticle diffusion   39

l

liquid molar volume   109

m

MATLAB
  modeling   156, 172
  optimization   218, 226
membrane refining   92
mixture separability   207
modeling
  artifical neural network (ANN)   53, 168, 172
  black box   53, 54, 155, 156, 159
  DOE   162, 166, 170, 175
  equation of continuity   50
  equilibrium constant   150
  gray box   53, 54, 154, 156, 158, 159, 161, 162
  mass balance   147–150
  mass transfer   44, 49, 147, 148, 149, 156, 195, 199, 223, 225
  statistical, ANOVA   167, 168, 172, 188
  velocity   53, 147–149
  white box   49, 53, 155, 156, 161
modifier   17, 39
molecular distillation   92

n

near critical region   3, 5
normal boiling point   109
number of stages, equilibrium stages   207
  column capital and operating costs   208
  optimum number of stages   208

o

optimization   73, 75, 76, 181, 214, 217, 222, 229, 231
  optimal operating conditions   212

overall extraction curves
  effects of parameters 47
  profile 43–45
  simulation 147

p

palm oil
  carotenoids (α-carotene & β-carotene) 84, 90
  constituents 84
  density of coexisting phases of CPO-CO_2 system 114
  distribution coefficient of CPO components 134–138
  endocarp 83
  exocarp 83
  free fatty acids (FFA) 84, 106
  kernel 84
  mesocarp 83
  minor components 84, 106
  phase equilibrium data 112
  predicted phase equilibrium of CPO-CO_2 system 133
  predicted physical properties of pure components 123
  recovery, refined palm oil 208–210
  solubility in supercritical CO_2, 22–24, 115, 206
  ternary system–CPO–triglycerides–FFA 133–134
  triolein 105
  tripalmitin 105
  vapour pressure 120
  vitamin E (tocopherols & tocotrienols) 90
palm oil or crude palm oil (CPO) 18, 83, 103
palm oil triglycerides-saturated fatty acids
  palmitic, stearic 84–88
palm oil triglycerides-unsaturated fatty acids
  oleic acid, linoleic acid 84–88
Pareto chart 187
particle diameter 154, 217, 225
Peng–Robinson equation of state 109
phase equilibrium modeling 101
product purity 205, 212
profit, optimum conditions 222–223
pure component critical properties 108
  critical temperature 111
  critical pressure 111

r

R134a
  properties 31–34
  reduced pressure 192, 193
  reduced temperature 192, 193
  solvating power 33
  subcritical 33–35
raffinate 11, 118
Redlich–Kwong–Aspen equation of state 102
reduced temperature 109
reflux ratio 209
Regression of binary interaction parameters 122

s

selectivity, K-factor 23
sensitivity analysis
  analysis of optimum point 223, 224
separation factor between CPO components 138–141
separation performance 205
simple countercurrent extraction (without reflux) 113, 118, 212
solid matrix 42
solubility-controlled phase 44

solvent extraction   92, 99
solvent–solute
  compatibility   20
  saturation   38
  equilibrium   38
solvent-to-feed (S/F) ratio   116, 208
steam refining (physical
    refining)   92, 97–99
  deacidification   93
  deodorization   93, 96
subcritical
  physical properties   3–7, 19
  polarity   7, 19
  R134a   33–35
subcritical extraction
  Orthosiphon (Java Tea)   168
  Orthosiphon stamineus, 231
supercritical CO_2
  carotenes recovery   18
  crossover   37
  decaffeination   13–15
  FFA removal   15–16
  oil and fats processing   15–16
  properties   3
  refining   92, 99
  solvating power   31–33
  tocopherol enrichment   17
supercritical extraction
  Anise   153–155
  diffusion-controlled phase   44
  dynamic method   24
  external mass transfer   39
  free oil   42
  intraparticle diffusion   39
  process   7–11

Q. infectoria, 164
solid matrix   42
solubility-controlled phase   44
static method   24
tied oil   42
supercritical extraction optimization
  chamomile   225
  nimbin   220, 221, 223
  Passiflora seed   153
  passion fruit seed   173, 176
  modeling mass transfer.   148, 149
supercritical fluids
  carbon dioxide   29–31
  depressurize   9
  history   1–2
  physical properties   3
  solubility   18
  solvent capacity   21–22
swelling   202, 203

t

theoretical stages   117
tied oil   42

v

vapour pressure   107
  experimental vapor pressure
    point   110
  reduced vapor pressure   109
volatility
  organic compounds   21, 23
  solutes   19, 23

x

xantine   11

Printed and bound by CPI Group (UK) Ltd, Croydon, CR0 4YY